Ebola: The Dreadful History

by

Mary Bradley-Cox

Contents

Preface

2013 was the year in which West Africa's worst outbreak of Ebola began. In December, a small, two-year-old child, Emile Oumouno, in the village of Meliandou in south east Guinea, became ill with a fever, diarrhoea, vomiting and intestinal bleeding. Within days he was dead, followed by his three-year-old sister Philomene, their mother Sia who was pregnant with a new sibling for them, and their grandmother. Two nurses who had been caring from them also died. All of them seemed to have died from the same disease, but nobody in Meliandou knew what it was. Emile Oumouno was Patent Zero at the beginning of a terrifying epidemic that has claimed nearly 9000 lives in countries from Guinea to the United States.

Where did the Ebola virus come from? How did it infect a tiny child? When did it first appear in Africa? Is it a man-made disease, as some have claimed, or has it been quietly existing in the west African forests for generations?

This book will trace the history of Ebola from its emergence in Zaire in 1972, to the epidemic that is continuing today in 2015.

The origins of Ebola

Ebola first emerged from the rain forests of Africa in 1972 when it infected a young missionary doctor, Thomas Cairns. Nobody knew what the illness was, and it was not to be identified as Ebola until 1976 when it broke out again.

The report sent by Dr Ngoi Mushola, the Bumba Zone Medical Director to the authorities in Kinshasa, about the first recognised outbreak of Ebola in 1976:

"The affliction is characterized by a high temperature around 39°C; frequent vomiting of black, digested blood, but of red blood in a few cases; diarrhoeal emissions initially sprinkled with blood, only now and then, retro sternal and abdominal pain and a state of stupor; prostration with heaviness in the joints; rapid evolution toward death after a period of about three days, from a state of general health."

Ebola and its close relative Marburg are members of the Filoviridae family, a group of viruses characterised by a linear, non-segmented, single-strand negative RNA genome. They are among the most lethal pathogens known to affect humans, and other animals, causing acute haemorraghic fever, and, once infected, often death within a matter of days.

Filoviruses are at least 16 - 23 million years old, arising at a time when the great apes were beginning their evolutionary journey. At one time, scientists believed that filoviruses evolved at the same time as agriculture, some 10,000 years ago, but new findings show that the ancestors of the Ebola and Marburg viruses are millions of years older than the development of agriculture by humans. Jeremy Bruenn has found traces of filoviruses in the Arctic, in rodents frozen 15 to 20 million years ago. Ebola and Marburg share a common ancestor, but viruses are constantly mutating, at the same rate as the influenza viruses, which makes them almost impossible to kill.

Viruses operate at the edge of life. They can do nothing on their own without invading, and aggressively occupying, a host cell. Many scientists believe that viruses originated when a piece of DNA or RNA broke off from a cell and learned how to spread to new cells. A virus may only have a dozen genes, but that can be enough to make it fatal to an animal or human with tens of thousands of genes, when it manages to invade just one of those cells.

Jo Revill has stated that, "DNA viruses are relatively stable, lumbering beasts, while the RNA viruses mutate fast, generating many new variants each time they hijack a cell. Each infected cell makes hundreds or thousands of viral offspring, each subtly different, and competing against its siblings for survival". Once an Ebola virus has invaded a host cell, it multiplies until the host cell's wall bursts, and thousands of new viruses explode in search of new cells.

Nobody knows when the Ebola virus first began to infect human beings. It has been suggested that the Plague of Athens, in which a quarter of the population of the Greek city died in 430 to 425 BCE, was caused by an outbreak of Ebola. The symptoms were described by Thucydides in the 'History of the Peloponnesian War': a severe fever, followed by bloodshot eyes, vomiting and bleeding, then bloody diarrhoea and skin lesions, then death. It certainly sounds like Ebola. In 1996, Dr Patrick Olson and his colleagues at the Naval Medical Centre in San Diego, argued that a Minoan fresco on the island of Santorini, depicts African green monkeys, known to carriers of the Ebola virus. It is conceivable that sick monkeys could have carried Ebola virus from Africa to Ancient Greece.

However, in 2006, the Greek archaeologist Effie Baziotopoulou-Valavani, uncovered a mass grave in Athens containing victims of the Plague of Athens. Greek scientists were able to extract DNA from the skeletons, and identified the pathogen that had killed them.

The cause of the Plague of Athens was not the Ebola virus, but typhoid fever. DNA extracted from the dental pulp of the teeth of victims revealed that they were killed by the bacterium Salmonella enterica serovar Typhi. The scientists tested for the presence of bubonic plague, typhus, anthrax, tuberculosis, cowpox and cat-scratch fever, without success. They also tested the soil in the mass grave to ensure that the remains had not been contaminated from the surrounding soil. These tests were all negative; the only positive results were for typhoid.

The Plague of Athens, now known to be caused by typhoid, occurred almost 3000 years ago; just a tiny blip in evolutionary time. Anthony Ramirez wrote in 1994, "It took eight million years or more for simian primates to change two per cent of their genetic structure to become human beings. By contrast, the polio virus can change two per cent of its genetic structure in five days - the time it takes to pass from the human mouth to the gut. Imagine what a virus could do with 2,400 years."

Although the Ebola virus has a genetic evolutionary story millions of years old, in its present form it is a new disease. It first appeared in 1972 in the African equatorial rain forest. There were reports that an unknown disease was killing people in Zaire, now the Democratic Republic of Congo. The disease was terrifying, killing 50 to 90 per cent of the people infected by it, and nobody knew what it was, or how to deal with it.

The Ebola family

Six distinct Ebola viruses have been identified, listed below in chronological order of outbreak:

• Ebola Zaire (identified in 1976)	1972
• Ebola Sudan	1976
• Ebola Reston	1989
• Tai Forest Ebola	1994
• Bundibugyo	2007
• Ebola Guinea	2014

Ebola Guinea, the virus responsible for the 2014 outbreak, has a thirty per cent different genetic makeup to its parent, Ebola Zaire.

To date (February 2015) there have been seventeen outbreaks of Ebola Zaire; eight of Ebola Sudan; seven of Ebola Reston; one of Tai Forest Ebola; and two of Bundibugyo Ebola, according to the World Health Organisation. Of the known Ebola viruses, Ebola Zaire is the most lethal, with a fatality rate of around 90 per cent.

There have been outbreaks of Ebola in different areas of Africa: Yambuku, Zaire in 1976; Nzara, Sudan in 1976; Tandala,

Zaire in 1977; Nzara, Sudan (at the same site as the 1976 outbreak) in 1976; Gabon in 1994; Cote d'Ivoire in 1996-97; Uganda in 2001-02; DR Congo in 2001-02 and 2002-03 (separate outbreaks); Sudan in 2004; DR Congo in 2007; Uganda in 2007-08; DR Congo in 2008-09; Uganda in 2011; Uganda in 2012; DR Congo in 2012; Uganda in 2012-13; and in 2014 in multiple countries in west Africa. This list does not include laboratory accidents or Ebola Reston.

Eric Leroy believes that each Ebola outbreak is random, caused by a local event like somebody eating an infected animal, or some fruit that has contaminated in some way. The Ebola virus is always present in reservoir species, and sometimes it crosses over and infects another species, like a gorilla, a chimpanzee, a duiker, or a human being.

On the other hand, Peter Walsh (an American ecologist who specialises in mathematical theories), believes that the Ebola virus is a newly emerged virus that first appeared in the Yambuku, Zaire, area, and has spread by infecting new populations of animals. Walsh and his colleagues suggest that all the known Ebola viruses today are descended from the one original ancestor, probably in Yambuku in 1976. Walsh, Leslie Real and Roman Biet illustrated their findings with geographical maps and family trees, demonstrating that there were strong correlations between the distance in miles from the Yambuku outbreak; and the distance in time from 1976; and the distance in genetic mutations from the possible ancestor. Their results show that Ebola Zaire from Yambuku, has spread across Africa.

Wittman and his colleagues isolated Ebola Zaire virus from the carcasses of great apes in the Gabon/Congo region, and discovered that the Ebola Zaire virus is able to mutate, and to recombine its genetic makeup very rapidly, which leads to new strains of the virus. Recombination and mutation enable a virus to emerge into the human population.

How does Ebola kill people? The virus can enter through the eyes, scratches on the skin, via the mucous membranes, or through having an injection with an infected needle. The virus finds a cell; binds to its surface, and proteins coating the virus allow it to enter the cell. Once inside, the Ebola virus begins multiplying, and leaves the cell when it is full and bursting with new virus particles. More and more cells are invaded and Ebola then attacks the body's immune cells that are trying to overwhelm it. Ebola switches off the release of interferon from the immune cells, which enables it to move through the body, invading the liver, spleen and lymph nodes. By this time the person will be feeling very ill, with fever, sore throat, muscle pains and weakness. The immune cells, that have become Ebola virus cells, then travel through the bloodstream, invading other organs. The cells release proteins that stick to the linings of blood vessels, causing clots, and eventually leaking.The normal blood flow to the kidneys and liver is blocked, leading to eventual shot-down. Because the liver cannot function properly any more, the blood loses its capacity to clot, leading to the bleeding seen in some Ebola victims.

A person infected with Ebola, bleeding from every orifice, as Richard Preston described the disease in 'The Hot Zone', "The skin bubbles up into a sea of tiny white blisters mixed with red spots known as a maculopapular rash...Spontaneous rips appear in the skin, and haemorrhagic blood pours from the rips...Your mouth bleeds, and you bleed around your teeth, and you may have haemorrhages from the salivary glands - literally every opening in the body bleeds, no matter how small." And at the end stage, "The internal organs, having been dead or partially dead for days, have already begun to dissolve, and a sort of shock-related meltdown occurs, and the corpse's connective tissue, skin and organs, already peppered with dead spots, heated by fever, and damaged by shock, begin to liquefy, and the fluids that leak from the corpse are saturated with Ebola-virus particles". Only a minority die like this. It is now known that the internal organs do not liquefy, but the lining of the blood vessels do break down, and leaking occurs from this. It is this leaking that causes blood clots all over the body, and leads to organ failure.

Eventually, fluid accumulates in the brain, leading to convulsions: this combined with multiple organ failure, leads to death. The whole of the body, by this time, is full of Ebola virus, which is why it is so dangerous to anybody touching it.

The search for the reservoir of Ebola

Where does the Ebola virus retreat to between outbreaks? In which animal species does it lurk, waiting for a primate or another species to infect?

A variety of animals have been identified by scientists as reservoirs. They include: baboons, macaques, chimpanzees, gorillas, duikers (small African antelopes), bats, green monkeys, and pigs, among others. Most people agree that the strongest candidate animal is the fruit bat. To be a reservoir species, the host must be able to tolerate the virus, without dying from the infection, and be able to spread it among likely victims.

Weingart and his colleagues found that, under laboratory conditions, Ebola could spread from pigs to macaques. Four macaques were placed in a cage within a pig-pen. Nose swabs were used to infect piglets with Ebola Zaire, and they were put into the pen. The macaques were in cages separated from the piglets by wire. The piglets developed mild fevers, and were breathing more heavily than usual, but recovered. The macaques all became infected with Ebola. Pigs easily acquire and spread infections, and in the case of the macaques, the virus was spread through the air as the pigs and the macaques did not directly come into contact with each other.

Duikers are a small antelope,and they have been found dead with other animals during human outbreaks of Ebola virus. Eating a duiker carcass could infect humans if it died of Ebola. Leroy found that there had been several outbreaks of human and animal Ebola in Gabon and the DR Congo between 2000 and 2004. There were simultaneous epidemics caused by different viral strains of Ebola, and these resulted from the handling of gorilla, chimpanzee and duiker carcasses.

Other scientists have also reported massive die-offs among gorillas and chimpanzees that coincide with human outbreaks of Ebola. Between 2002 and 2003, Ebola Zaire killed 5000 gorillas in and around the Lossi Sanctuary in North West Congo. It is thought that the infections were from the virus being passed from one group of gorillas to another, rather than spillovers from a reservoir host.

Formenty found that a quarter of the wild chimpanzees in the Tai National Park, Cote d'Ivoire, died of a new strain of Ebola, and stated that, "A high mortality rate among apes tends to indicate that they are not the reservoir for the disease causing the illness".

Chimpanzees have eaten humans, occasionally taking small children, particularly if an infant has been left near the edge of the forest while the mother has worked on her crops. In a similar way, humans hunt chimpanzees and eat them. Humans live side-by-side with wild animals, and constantly exchange bacteria and viruses.

Baboons also become infected with Ebola. They share 96% genetic homology with humans, indicating a shared ancestry in the evolutionary past. Cross-infection can occur through direct contact, or through breaks in the skin if a baboon is being handled, dead or alive. It is easy for the skin to be cut or nicked while butchering an animal carcass.

Bats have been identified as the main reservoir of Ebola virus, after thousands of insects, reptiles, and other animals have been tested as carriers of Ebola by numerous scientists. Ebola Zaire antibodies were found in Hammer-headed bats; Epauletted Fruit Bats; and Little Collared Fruit Bats during outbreaks in Gabon and the DR Congo between 2001 and 2005.

Marburg virus antibodies were also found in Egyptian Fruit Bats in Kitaka Mine in Uganda in 2007. Ebola Zaire antibodies were found in a single migratory fruit bat in 2008 in Accra, Ghana. This bat is common in sub-Saharan Africa, and lives in large colonies, often in cities. The flight of an individual Eidolon helvum fruit bat during migration has been recorded as more than 2500 kilometres, which is worrying if the virus is found to be spread by the species. Fruit bats were tested for antibodies for Ebola Zaire and Ebola Reston in Bangladesh in 2012. Alarmingly, 3.5% of bats out of 276 tested were positive. This could be a reservoir for Ebola viruses in Bangladesh, and could extend the range of Ebola into Asia. Orangutans in Indonesia were tested and some had positive antibodies to Ebola virus. The WHO stated that this suggests, " the existence of multiple species of filoviruses or unknown filovirus-related viruses in Indonesia, some of which are serologically similar to African ebolaviruses."

Scientists during the 1976 Ebola Sudan outbreak in Nzara, scrutinised the cotton factory where the index case had worked. Rats were found in every room of the factory, and bats were roosting on the ceilings. Rats, bats, birds, lizards and insects were trapped in an effort to find the reservoir of Ebola in this outbreak. The bats were found to be 'forest-dwelling' bats - there because the town of Nzara had not existed until quite recently; it had been constructed on cleared rain forest land. No Ebola virus was found in any of the wildlife trapped in and around Nzara.

In Zaire, it was found that Ebola had spread to 55 villages around Yambuku. A total of 318 people were infected with Ebola, and 280 had died, a lethality of 90%. People had contracted the virus though injections at the Yambuku Mission hospital, but the index case was a man called Mabalo Lokalo. He had been on a journey for twelve days, a holiday with other local men; on the way home they had stopped at a roadside market and bought freshly killed antelope, while one of his companions bought a dead monkey. When they returned home, the animals were cooked and eaten. Mabalo Lokalo's family were sure that he had only eaten antelope, and none of the monkey. Perhaps the antelope was infected, or perhaps just handling the dead animals was enough to pass the Ebola virus to Mabalo Lokela

In 1996 in Mayibout 2, Gabon, near the border with Republic of Congo, a group of villagers took part in the butchering and eating of a chimpanzee found dead in the forest. All the people involved became infected with Ebola, and 21 out of a total of 37 died.

Most scientists seem to agree that the reservoir for Ebola is in bats, but how does it get into the other animals? Where the bats roost, their droppings fall onto surfaces that will probably be touched by humans in buildings; and in the forest droppings fall onto fruits that are consumed by monkeys, chimpanzees, gorillas, antelopes and other animals. The first indication of an Ebola outbreak may be when there are animal deaths in the forest, more than would be usual.

In the past five years (to February 2015) there have been five Ebola outbreaks affecting humans in West Africa, that have killed thousands of people. Scientists believe that hundreds or even thousands of animals have also died from Ebola. Gorillas, chimpanzees and duikers have all been affected, and in some areas their numbers have not recovered. Where there used to be hundreds of great apes, there are now none. The first warning signs of an Ebola outbreak can be animal carcasses in the forest, and local people should avoid handling, butchering, cooking and eating the dead animals as they may be infected with Ebola.

The 1970s - Ebola begins

There was an unreported case of Ebola in 1972, perhaps the first time that a non-African has survived an infection with the virus. Dr Thomas Cairns, an American, was working as a missionary doctor at a hospital in Zaire. A patient was admitted with a severe fever, so ill that he was almost dead. Within minutes he had died, and the hospital asked Dr Cairns to perform an autopsy so that they would know what disease had killed him. While performing the autopsy, Dr Cairns nicked his finger with a scalpel. Two weeks later he became ill with what at first seemed like influenza. He describes it as, "the worst kind of flu you can imagine, aching all over", with a severe headache, vomiting and diarrhoea. Most of the time he was ill, Tom Cairns was unaware of anything, apart from feeling incredibly ill. It took six weeks for him to recover, and he had lost 20 pounds in weight, and did not regain full health for months.

Dr Cairns was still working in Zaire in 1976, when two epidemics of a serious, unknown virus broke out in northern Zaire and southern Sudan. There were 602 cases, with 431 deaths. The Centers fro Disease Control began testing health workers in the area to determine whether any of them had been infected with he mystery virus. All 50 were negative, apart from Dr Cairns who had antibodies for Ebola virus in his blood. He finally had a diagnosis of his illness: Ebola. He was the first non-African survivor of this terrible disease. Since then his blood has been studied at the CDC in Atlanta; it has been found that the number of antibodies in his blood have reduced over time - perhaps eventually survivors of Ebola lose their immunity to the virus?

Nobody knows when the Ebola virus first encountered humans. The first reported cases were in 1976 in what was Zaire, now the Democratic Republic of Congo, and in Sudan.

The outbreak was in the Bumba region of Zaire, centred on the Yambuku Mission hospital. The unknown disease rapidly became worse in those infected, with a high fever, muscle and joint pains, sore throats, red eyes, and vomiting and diarrhoea. There was bleeding from the nose and gums, sometimes sites of injections, and also massive haemorrhages. Medical teams knew that they were dealing with a haemorrhagic fever, but they did not know if it was yellow fever, or Lassa fever, Crimean Congo, or even Marburg virus.

The index case was a schoolteacher named Mabalo Lokela, who went to the mission hospital with a high fever. The nuns thought that he had malaria, and gave him an injection of quinine, and sent him home. Two days later, Mabalo Lokela was extremely ill; he had a headache and was dizzy; he had trouble breathing; he was vomiting and had diarrhoea; and then he began bleeding from his gums, and from his intestine. Two weeks after he became ill, ge died. Mabalo Lokela's wife, his mother and his sister prepared his body for his funeral according to their custom. Within days, they too had the haemorrhagic fever.

At the same time as the Yambuku outbreak, a similar haemorrhagic fever appeared in Nzara, in south Sudan. The authorities sent blood and tissue samples from victims of both outbreaks to the CDC in Atlanta, USA; the Institute of Tropical Medicine in Antwerp, Belgium; the Microbiological Research establishment at Porton Down, in the UK; and to the Pasteur Institute in France, hoping that one of the laboratories would be able to determine what the illness was. Over half of the infected patients in Nzara died, and a terrifying ninety per cent of the victims in Zaire.

Peter Piot was working at the Institute of Tropical medicine in Antwerp, Belgium, when a thermos flask arrived from Zaire. The flask contained vials of blood, one of them broken in transit from Africa. They were samples from a nun working at the Catholic mission in Yambuku, who had become ill with the mysterious illness. Piot, a clinical biologist, and his colleagues did not take any special precautions with the samples. They put a slide with a blood sample under a microscope; what they saw when they looked down the microscope shocked them. "We saw a gigantic, worm-like structure - gigantic by viral standards. It's a very unusual shape for a virus; only one other looked like that, and that was the Marburg virus", said Piot.

In 1967, people were infected with a haemorrhagic fever in Marburg, Germany, and also in Belgrade. The virus they were infected with was unknown to science, and it was named Marburg. Viruses are traditionally named after the place where they are first discovered.

Piot conferred with his colleagues, and they realised that this new virus was similar to Marburg, but was something never seen before. The Institute was informed that the sick nun had died, and that numerous other people were sick and dying. Piot and a team decided to go to Yambuku to investigate this mysterious virus.

The leader of Zaire, President Mobutu, arranged for a transport aircraft to take the Belgian scientists as close to Yambuku as they could. The Bumba Zone was quarantined. All roads, waterways and airports were now under martial law in an attempt to stop the new disease from spreading. The aircraft carrying the scientists landed in Bumba, the closest airfield, and the pilots agreed to stay on the ground long enough for Piot and his team to disembark their equipment; and then rapidly took off again. Everyone was terrified of catching the new disease. The Catholic mission at Yambuku was 75 miles from Bumba, and to reach it meant travelling along red earth roads with deep holes and ruts from which the vehicles would have to be dug out time and time again. The convoy finally reached the mission to be confronted by a card with a sign across the doorway of a building where a group of nuns and a priest were waiting. The sign said, in the local Lingala language, "Please stop, anybody who crosses here may die".

The scientists had to know how the virus

was transmitted from person to person. Was it through the air, in food, by insect bites, or by touching? The surveillance teams went out to find any cases of Ebola, as the virus had been named, after the Ebola river nearby. Maps were drawn, and each village visited was plotted on the maps. Every infected person was logged, with where they lived. It was discovered that all ages and both sexes had become infected with the virus, but women aged 15 to 29 had the highest incidence. Many of the women had attended the ante-natal clinic at the mission, and all had received an injection of vitamins. The scientists discovered that every morning just five syringes were prepared at the hospital, and used again and again: this was the way in which the virus had been transmitted, unknowingly by the nuns, between victims.

Many of the villages around Yambuku had taken their own precautions against the disease. They had erected roadblocks on the roads going in and out of the village, with men from the village stopping anybody from entering or leaving. Anybody who was sick, and their families, were kept apart from the rest of the village in a hut. Anybody who died was buried well apart from the houses. The village elders knew what epidemics could do; they remembered outbreaks of smallpox, and how to deal with them. Unfortunately, this new disease did not behave like anything they had seen before.

The surveillance team then discovered that people were becoming ill after attending the funerals of people who had died of the virus. It is the practice in Zaire for a dead person's body to be washed and handled by the women of the family, and prepared for burial. The body of an Ebola victim is literally full of virus - touching it can be lethal. When patients died , the medical team buried the body without allowing any contact at all with grieving relatives. This was extremely upsetting for those mourning loved ones, but it was the only way to prevent them becoming infected themselves.

On 26 November, an American Peace Corps volunteer named Del Conn became ill in Yambuku. He had a severe head and back ache. By the 29th he was much worse, with nausea and Ebola symptoms. The Zairean helicopter pilots who were called in refused to take him to Bumba, so the poor man was put in the back of a Land Rover and driven the four and a half hours over bumpy, pot-holed roads. Everybody in the Land Rover wore surgical masks and protective clothing. When they arrived at Bumba, people were terrified, and the Land Rover was marooned on the airport landing strip as the pilots refused to carry the sick man, and they could not take him to the town itself. The group from Yambuku had to spend the night in the land Rover. Fortunately, the next morning an Air Zaire aircraft arrived with a Canadian negative-pressure isolation unit. The sick man was flown to Kinshasa, to find that no aircraft was available to take them on to South Africa. The group spent another uncomfortable night in makeshift surroundings in a deserted hangar. The US government sent a US Air Force aircraft the next morning, with an Apollo space capsule to keep Del Conn isolated in Johannesburg until he could be sent home to the US. The plan for his evacuation that had been conceived in Yambuku, planning a 34 hour journey to get to Johannesburg eventually took twice that time. When Conn's blood was analysed, he was found not to be suffering from Ebola, but from another unknown viral infection.

The Yambuku outbreak lasted from 1 September to 24 October 1976. 318 people were infected with Ebola, and of those 280 died. There were only 38 known survivors.

The scientists were unable to establish how the Ebola virus spread from Nzara in Sudan, to Yambuku in Zaire. It is possible that a person travelled from one are to the other, became ill and was treated with an injection, the needle then being used on somebody else, thus spreading the virus.

While the 1976 Yambuku outbreak of Ebola was raging, there were reports of a similar illness occurring in southern Sudan. Yusia G. Was the first person - patient zero of the epidemic in Sudan. He worked tallying the bales of cotton manufactured from the local cotton at a factory in Nzara. On 27 June 1976, Yusia became ill with a headache, and then a very sore throat. This was followed by sever muscle pains in his chest, neck and back. On 30 June, Yusia's brother Yasona was so alarmed at his brother's condition that he took him to the hospital in Nzara. Yusia was by then also suffering from abdominal pains, diarrhoea and vomiting. On 2 July he began to bleed from his nose and mouth, and the continuing diarrhoea was full of blood. On 6 July, Yusia died. On 13 July, Yasona became ill, but he survived. In total, 48 people developed Ebola, who worked, or had contact with, people who worked at the cotton factory. 27 of the 48 died.

The World Health Organisation decided to send a team to investigate the Sudan outbreak, led by David Simpson. With him would go David Smith from the Medical Research Council unit in Nairobi; Don Francis, an epidemiologist; and entomologist Barney Houghton. They were to look for an animal reservoir for Ebola. No provision had been made for the investigative team. Dr Giocometti, an Italian WHO representative gave them a Land Rover and a driver, and the Juba District Ministry of health donated some tyres and 100 gallons of petrol. After hours of slow driving on an unmade road, they finally arrived at the hospital at Maridi that they thought would be treating Ebola victims. Most of the hospital was deserted. One of the workers, Samir, from the cotton factory at Nzara, had been brought to the hospital by his family as he was ill with a fever. Samir's arrival had triggered an explosive epidemic within the hospital. Fifty per cent of the ward staff were already dead, including the only doctor at the hospital. The survivors had deserted the hospital, and had fled into the bush, hoping that the virus would not find them.

The Sudanese government had isolated the south of the country. No aircraft, cars, or any other type of vehicle was allowed to enter or leave the exclusion zone. It caused severe food shortages in the south, but the Sudanese government was determined that the disease would not infect the north, with its mainly Muslim population.

The only method that scientists could use in 1976 to determine the nature of a virus was to look for specific antibodies in the blood of infected patients. Another method was to inoculate animals like guinea pigs, or mice, in the laboratory, then observe to see if they sickened, then conduct autopsies on the bodies, if they died, to determine what had killed them. Or scientists could breed the virus in cell cultures grown in flasks, and then observe what resulted under microscopes.

When a virus as lethal as Ebola seemed to be, an ordinary laboratory is not equipped to deal with it. Samples from the outbreaks in both Zaire and Sudan were sent to the CDC in Atlanta, and to Porton Down in the UK.

On 5 November, a scientist at Porton Down accidentally injected himself with Ebola virus while injecting a guinea pig. The virus had been with the virologists since 5 October, and they thought they might be dealing with Marburg. Guinea pigs had been inoculated with serum from the infected nuns in Zaire, and all the animals had died.

Meanwhile, scientists at the CDC in Atlanta used serum from people who had survived Ebola. The serum was tested for Marburg, but had proved negative. A new haemorrhagic virus had been identified, and named Ebola, after the Ebola river that flowed across northern Zaire.

The scientist at Porton Down, Geoffrey Platt, became ill a week after the accidental needle-stick. He was moved to Coppetts Wood Hospital in North London under high security. All the other patients in the hospital were removed, and sent to other hospitals for their safety. Nobody knew how Ebola transmitted; it could have been an airborne virus, so all precautions were taken. Platt's ward was completely sealed off in a quarantine capsule. Porton Down was closed, and anybody that Geoffrey Platt had come into contact with was quarantined. Platt developed Ebola, but recovered. His ward was fumigated so thoroughly that even the isolation tent was burned and buried, to avoid the slightest risk of contamination. Geoffrey Platt lost a great deal of weight, and suffered desquamation while recovering, losing his skin and hair. The virus is so toxic, that many people lose their hair, fingernails and skin. Those who survive grow new skin, hair and nails.

There were two other outbreaks of Ebola in the 1970s. In Tandala, Zaire in 1977, a nine-year-old girl died of acute haemorrhagic fever in the hospital; this was the first recognised case of Ebola after the virus had been named in 1976. Then in 1979, Ebola Sudan struck again at the cotton factory in Nzara, Sudan. In August, a man who worked at the cotton factory was admitted to the hospital with fever, diarrhoea and vomiting; he died three days later. Three of the man's relatives then developed Ebola, as did another family. Two nurses who had looked after the Ebola patients also died, and the authorities realised that an Ebola epidemic was underway. 34 people were infected, with a 65% fatality rate before the outbreak ended.

Every person infected had a link to the Nzara cotton factory where the first epidemic in 1976 had occurred.

After these outbreaks into the human population, Ebola Zaire and Ebola Sudan disappeared back into the rain forests, and did not strike again until the end of the 1980s.

Ebola in the 1980s

Ebola suddenly reappeared in the United States in Reston, Virginia in 1989. Animals began to die in a monkey facility in Reston, bleeding from their noses. The Hazelton facility imported crab-eating macaques from the Philippines, and at first the scientists thought they were dealing with an outbreak of Simian Haemorraghic Fever, that had previously only been seen in Africa and India. They thought that perhaps the monkey importers had smuggled in some that were sick from India or Africa. The laboratories at the United States Army Medical Research Institute of Infectious Diseases, Fort Detrick in Maryland were not too far away, and the scientists there were asked to investigate the outbreak.

Specimens taken from the sick monkeys at the Hazelton facility at Reston were fascinating to the scientists. Cells in one of the flasks had been completely destroyed, and one of the scientists even sniffed the flask to see if there was any bacterial contamination, but it had no odour. No particular precautions were taken to guard against infection until the organs from the dead monkeys were examined; the monkeys had died of internal haemorrhages, a trademark of Ebola. As one-by-one the monkeys died, they realised that they had an epidemic an their hands, and the workers at the facility who came into contact with the monkeys were watched to see if any of them became sick. On 4 December, one of the workers became ill, had a fever and was vomiting. The man was immediately rushed to hospital and put into isolation. Within 24 hours it was confirmed that the man did not have Ebola.

The fear of any people contracting Ebola was such that it was decided to euthanase the remaining monkeys. In total 450 of them were put down, with one trying to escape and leading the men trying to catch them a hair-raising chase around the building. The people working there thought that one bite might lead to Ebola, so everyone was afraid. Four of the men did catch the virus, but it was found that Ebola Reston was fatal to monkeys, but just caused a mild fever in humans. The only worrying thing about the outbreak was that Ebola Reston was transmitted through the air. Monkeys in cages in one room infected monkeys in cages in another room down a corridor. It was thought that infection may also have occurred via the air ducts.

Tracing the outbreak back to the Philippines, it was found that there were four main dealers in Manila, the capital. The animals had been shipped via KLM flight that stopped-over in Amsterdam en route to Virginia. The monkeys only stayed for six hours in Amsterdam. The scientists were trying to identify the source of the Ebola Reston virus - was it in the Philippines, or had it been in Amsterdam? The men from USAMRIID were then told that more monkeys were travelling to the US. These were being shipped from Arusha in Tanzania to texas, stopping again in Amsterdam, the hub for KLM, the Dutch airline.

Exploring every possibility for a monkey to have become infected with Ebola on its way to Virginia, or possibly en route to Texas, the scientists scrutinised any monkey that had shared any space with the Reston monkeys. Some from another flight had shared a room
at the animal holding facility in Amsterdam en route to Mexico City. Some of those monkeys had antibodies to Ebola, indicating that they too had had contact with the virus.

At JFK airport, scientists decided to examine the animal hostel run by the ASPA (American Society for the Protection of Animals), for any monkeys flying via the airport. They were horrified to learn that a volunteer at the hostel had been sick two years previously with what sounded like a haemorrhagic fever. When the volunteer was tested, she was found to have a low level of antibodies for Ebola in her blood. It was impossible to determine whether or not she had been infected with Ebola Reston.

There was another outbreak of Ebola Reston at the Hazelton Laboratories facility in Alice, Texas, at the end of January 1990. From the beginning of February until the middle of March, monkeys were dying from Ebola Reston. The monkeys had been transported from the same facility in the Philippines. Four of the people who looked after the monkeys in Texas were tested and found to have antibodies for Ebola Reston, but they did not develop any symptoms.

Another shipment of monkeys was sent from the same facility in the Philippines to Philadelphia. Monkeys became sick and started to die in the same manner as those at Reston, Virginia.

The outbreaks of Ebola Reston in the US prompted the Philippines government to undertake its own investigation of the monkey facility in Manila: they were concerned that people working at Ferlite

Farms, where the wild monkeys were held before being shipped overseas, would be contaminated with Ebola. 186 people were examined, 48 were from wildlife collection teams, with the remaining 138 being from four monkey export facilities in the area (Ferlite Farms being one of them). Twelve of the people tested had antibodies for Ebola Reston, and 22% of of the workers at Ferlite Farms also had antibodies in their blood. All the monkeys at Ferlite Farms were put down to prevent further infection. Ferlite Farms continued their operations with fresh supplies of wild monkeys, and some were sent to Sienna in Italy in 1992, and they also became sick and died of Ebola Reston.

In 1996, there was yet another outbreak of Ebola Reston at the Hazelton Primate centre in Alice, Texas. Unbelievably, Hazelton had also reopened their Reston centre, and yet more monkeys were dying there. In the light of all these experiences, it would seem to be more sensible to obtain monkeys from a different supplier, or to breed monkeys in the US, rather than overseas, especially in the Philippines. Monkeys were needed for experimentation as the US military was researching the use of Ebola virus as a military biological weapon, for defence purposes of course.

Ebola in the 1990s

From its emergence in Zaire in the 1970s, extending to reach across Africa to Sudan, the US, the Philippines, Italy and the UK, the Ebola virus began to appear more frequently. During the 1990s, there were outbreaks in Gabon, Cote d'Ivoire, South Africa, and Uganda. A new strain of Ebola, Bundibugyo, emerged in the Democratic Republic of Congo.

Scientific expeditions were sent to the rain forests of Africa to try and locate the reservoir of Ebola, but nothing was ever found. People were tested, and their blood taken to look for antibodies. Amazingly, in the Central African Republic, 20 to 30% of people tested were positive for Ebola antibodies. There may have ben many more people with Ebola: if a patient presented at a hospital with a fever and diarrhoea, it could be any one of a number of tropical diseases, including yellow fever and malaria.

In December 1994, there were outbreaks of Ebola in the gold-mining camps deep in the rain forest. There were reports of dead chimpanzees and gorillas in the forest, and of the gold-miners butchering and eating the carcasses. The sick miners were taken by boat to Makokou hospital, where doctors at first thought they were sick with yellow fever. People around the hospital began to fall ill. A traditional healer, a nganga, lived near one of the villages affected by Ebola, and people went to him for help. They preferred traditional medicine rather than the hospital, as sick people went there for help, and died. Some sick people stayed in the nganga's home while he was treating them, and they became infected with Ebola; also infected were families of people sick with Ebola in the hospital It is the tradition in Africa for families to prepare food, and look after their relatives when they are in hospital. By February 1995, there had been 52 cases of Ebola, of whom 31 died, a fatality rate of 60%.

In 1994, a scientist was infected with Ebola after conduction an autopsy on two chimpanzees that had been found dead in the Tai Forest, Cote d'Ivoire. The scientist wore gloves, but no face mask. Here gloves were not cut or pierced during the autopsies, but she became ill within days. Nobody knows how she was infected. She was flown to Switzerland for treatment; nobody suspected Ebola as it had not been seen before in Cote d'Ivoire, and she tested negative for Ebola Zaire and Ebola Sudan.

However, she did have antibodies for a new strain of Ebola; it was named Tai Forest Ebola virus.

Ebola struck again in 1995 in Kikwit, Republic of Congo. Cases began arriving at the hospital, and there was panic among the population, once they realised what the sickness was. The army erected barricades on all the roads leading out of Kikwit in an effort to contain the outbreak, and sent to the World Health Organisation for urgent assistance. Scientists were dispatched to work to stop the outbreak.

The index case was a farmer who had cleared some land in the forest to grow crops to feed his family; he also made charcoal. He became ill and died, infecting several members of his family who had cared for him. One of the family infected a laboratory technician named Kinfumu who became very sick, and went to the hospital in Kikwit for help. He had a high fever and a distended abdomen and, at first, the doctors thought that he had typhoid.; but the stomach distension indicated a perforated intestine. An urgent operation was carried out, and surgeons removed Kinfumu's appendix that appeared diseased. Kinfumu's abdomen continued to be distended, and the surgeons operated again the next day; they discovered that the abdominal cavity was full of blood. Kinfumu's blood went all over the operating theatre. Kinfumu died, as did all those who had come into contact with the blood. By the end of the outbreak, 315 people had become infected with Ebola Zaire, of whom 250 had died; a fatality rate of 81%.

The following year, 1996, Ebola surfaced again in Gabon. A hunting party had set out from the village of Mayibout 2. They found a dead chimpanzee in the forest, and took it back to the village to be butchered, and cooked in the local style. An African sauce was prepared, made of peanut gravy, with pili-pili (hot peppers) poured over fufu (made from manioc flour). Even though the chimpanzee had been a little decomposed, the hot sauce would hide the flavour. The hunters reported seeing French soldiers in the forest; they had arrived in a Zodiac boat, and seemed to be hunting as they heard shots. It was after the soldiers left that the men from the village had found the chimpanzee, and some of the villagers suspected that the soldiers had used poison. A total of 37 people became ill in Mayibout 2, and 21 died.

There was another outbreak in Gabon in

1995. A hunter in the Booue area found a dead chimpanzee and ate it. People who came into contact with him also died, and in total 60 people were infected with Ebola, of whom 45 died. A South African health worker who had helped treat the sick became ill while flying back to Johannesburg, and went into hospital. He survived, but a nurse who looked after him became infected with Ebola and died.

Ebola Reston also broke out again in 1996. It was among crab-eating macaques from the same facility, Ferlite Farms in the Philippines., from the outbreak in 1989. The same US company, Hazelton Laboratories, was involved in importing the monkeys into the US, to the same centre in Alice, Texas. No human contamination was involved.

After the series of outbreaks, caused by infected animals from the same facility in the Philippines, perhaps it would have been better if Hazelton had found another supplier.

Ebola outbreaks 2000-2014

Since Ebola virus first emerged from the rain forest and infected humans in the 1970s, the frequency of the outbreaks has escalated.

There were five outbreaks during the 1970s: Yambuku, Zaire; Nzara, Sudan; one scientist in England with a needle-stick; one retrospective case at Tandala, Zaire; and another at Nzara, Sudan.

The Reston Ebola virus emerged in the US in 1989, and in the Philippines. There were no other known outbreaks during the 1980s.

During the 1990s, Ebola emerged sporadically. There were outbreaks in Gabon; Uganda; separate occurrences of Ebola Reston in the US; a laboratory worker in Russia with a needle-stick, who died; and a health worker on returning home to South Africa, who infected a nurse. The health worker survived, but the nurse died.

The new millennium saw increasing outbreaks of Ebola. In October 2000, Ebola Sudan broke out in the Gulu District of Uganda. The authorities in Uganda moved swiftly to contain the virus. Samples were sent from infected people to the National Institute of Virology In Johannesburg, South Africa, who confirmed that the haemorraghic fever was indeed Ebola Sudan virus. The Ministry of Health in Kampala, Uganda, set up surveillance and control of suspected cases of Ebola, and a field laboratory was established at st Mary's hospital by the CDC, to allow faster diagnosis, which was supplemented with additional testing at the CDC in Atlanta, and in Johannesburg.

Burial teams were trained to deal with the Ebola victims - it is difficult for people to lose a loved one, and not to be able to bury him or her without the customary funeral practices, but local people were educated in how Ebola was transmitted through bodily fluids, so handling a corpse brimming with Ebola virus was a sure way to catch the disease. The people were also taught that contact with family members with Ebola could only take place with adequate safety measures. Taking care of a sick person was also fraught with danger if gloves and masks were not used, and bleach or chlorine to wash the hands, In total, 425 people were infected with Ebola, and of those 224 died, a fatality rate of 52%.

The next outbreak was of Ebola Zaire that

occurred over the border of Gabon and the Democratic Republic of Congo, between October 2001 and July 2002. By the end of the outbreak, 122 people had been infected, of whom 96 had died, a fatality rate of 90%.

In December 2002, in Mbomo, Republic of Congo, a group of hunters handled dead gorilla and duiker carcasses, and became ill after butchering and eating them. The illness spread to Kelle, and by April 2003, 143 people had been infected, of whom 128 died, a fatality rate of 90%

Ebola Zaire returned to Mbomo in November 2003, killing another 29 people before vanishing again into the rain forest.
In 2004, Ebola Sudan reappeared in Yambio in Sudan, 1500 kilometres from Mbomo. 17 people were infected, of whom 7 died.

Ebola returned again to Mbomo in April 2005, infecting 12 and killing 10 of them, a fatality rate of 83%. Why did Ebola keep appearing in Mbomo? Various theories have been pur forward by scientists. One theory is that the virus is always present in the reservoir species, and an animal like a chimpanzee eats some fruit that has been defecated on by the reservoir, perhaps a fruit bat; develops Ebola and dies. A human finds the dead chimpanzee, and takes it back to a village to be shared and another outbreak occurs. Another view is that Ebola is a new virus that is expanding its range with each new outbreak, which would explain the variety of locations.

Ebola returned to the Republic of Congo in 2007, to the province of Kasai Occidental. This was a very large outbreak, with 264 people infected. Laboratory tests conducted by Institut National de Recherches Biologiques (INRB) in Kinshasa on urine and blood samples collected from suspected cases, have also confirmed the presence of Shigella dysenteriae type 1, further complicating operations while case definitions and clinical descriptions, particularly in response to rehydration and antibiotic treatment, support a possible concurrent outbreak of another etiology. A enhanced team of national and international experts is being mobilized to implement control strategies for Ebola haemorrhagic fever and to support outbreak field response in the province. Médecins sans Frontières (Belgium) has deployed clinicians, water and sanitation experts and logisticians to the area and has established appropriate isolation facilities. The WHO were supporting the Ministry of Health in Kinshasa and in the field at the location of the outbreak. Additional staff, outbreak response equipment and supplies, including Personal Protective Equipment (PPE) are being sent to the area. A enhanced team of national and international experts is being mobilized to implement control strategies for Ebola haemorrhagic fever and to support outbreak field response in the province. Médecins sans Frontières (Belgium) has deployed clinicians, water and sanitation experts and logisticians to the area and has established appropriate isolation facilities. Of the 264 people who were infected, 187 died, a fatality rate of 71%.

Barely a month later, a new strain of Ebola emerged in Uganda. Twenty deaths were reported in Bundibugyo to the Ministry of Health on 5 November. The symptoms were fever, fatigue, headache, vomiting and diarrhoea, muscle and joint pain and abdominal pain. Ebola virus was suspected, and samples were immediately taken and sent to the CDC in Atlanta, US. On 28 November the CDC announced that the disease was caused by a new strain of Ebola, named Bundibugyo. The new virus was genetically 32% different to the other Ebola viruses, suggesting that the virus had evolved with, perhaps, a different reservoir host. The people in Bundibugyo reported that a monkey had bitten a goat, which had subsequently died. The goat was butchered and cooked and eaten by a family, who then became ill. Not everybody became ill in the family, and the Bundibugyo virus did not appear to be as lethal as its relatives, Ebola Zaire and Ebola Sudan. The outbreak was over by January 2008; 149 people were infected, of whom 14 died, a fatality rate of 25%.

The Kasai-Occidental Province of Republic of Congo was visited again by Ebola Zaire in December 2008. 32 people were infected, of whom 17 died, a fatality rate of 17%. There was a lull of three years before Ebola struck again, between June and August 2012, in Kibaale, Uganda. 24 were infected, of whom 17 died, a fatality rate of 71%.

At the same time as the Kibaale outbreak, Ebola Bundibugyo returned in Province Orientale of the Republic of Congo. This time the virus was more lethal: 77 people were infected, of whom 36 died, a fatality rate of 47%.

Ebola Sudan virus surfaced in Uganda in November 2012, infecting 6 people, of whom 3 died. The outbreak ended in January 2013.

In March 2014, the outbreak of Ebola Zaire began in Gueckedou, Guinea, and it is still ongoing at the time of writing in February 2015. Over 21,000 people have been infected in a number of countries, and over 22000 infected, of whom nearly 9200 have died.

Scientists still do not know where Ebola hides between outbreaks, or what the reservoir is, although fruit bats are strongly suspected. People have been told not to touch, butcher or eat dead animals found in the forest, but between outbreaks, memories fade, especially when people are hungry.

The 2014 outbreak

The most widespread and deadly Ebola virus outbreak began in December 2013 with the death of a two-year-old little bay named Emile Ouamouno. Emile lived in the village of Meliandou in Guinea, with his parents Etienne and Sia; his three-year-old sister Philomene, and his extended family. Poor little Emile developed a fever, a headache and bloody diarrhoea. His grandmother and mother nursed him and did their best to make him better, but four days later he died. His death was followed by that of his sister. His mother, and his grandmother, who had nursed them all.

Nobody knows how Emile contracted Ebola. Perhaps he ate a piece of fruit that had been contaminated by the droppings of fruit bats, drawn to the area where he lived to feed on the oil palms. Perhaps, as has been suggested, he played inside a hollow tree where fruit bats roosted. However he became infected, the Ebola virus began to spread out of the village of Meliandou, and wide into Guinea.

A retrospective investigation by WHO revealed that the country's first case was a woman who was a guest at the home of the index case in Meliandou, Guinea. When the host family became ill, she travelled back to her home in Sierra Leone and died there shortly after her return in early January. However, that death was neither investigated nor reported at the time.

Meliandou is in the Gueckedou Prefecture in southern Guinea, near the borders with Sierra Leone and Liberia. It is 191 kilometres from Kenema, a journey of three and a half hours. Kenema and Kailahan in Sierra Leone, then became the epicentres of the Ebola outbbreak. Kailahun and, to its south, the larger city of Kenema, formed the early epicentre of the outbreak. WHO and other partners concentrated their response teams in that area.

Ebola killed village health workers in Meliandou who cared for the people first infected, not realising what they were dealing with. Ebola quickly spread into neighbouring villages. People are able to easily cross the borders into Sierra Leone and Liberia, spreading the Ebola virus beyond the Guinea borders before health authorities realised what was happening.

Kenema has a government-run hospital with a laboratory run by Americans to diagnose Lassa fever, which is endemic in the area. The laboratory diagnosed the first

Ebola case, but the isolation wards were soon full of patients with the disease. Eight of the nurses became infected with Ebola, and the hospital could not cope with the number of cases of the virus.

The Ebola epidemic was first reported in Guinea in March 2014, by which time hundreds of people were infected. The response from the World Health Organisaton was, sadly, lacking. An internal report was leaked from the WHO in October, that blamed incompetent staff and bungling bureaucracy for failing to spot the chances to stop Ebola from spreading. The WHO's Africa office is the United Nations agency that should have realised what was happening, but did nothing. It was left to private charities like Medicins sans Frontieres and Oxfam to fight the virus without any help from the rest of the world, who had not been informed that an incurable epidemic was on the loose in West Africa.

In April 2014, Medicins sans Frontieres told the WHO and the UN that the epidemic would be huge. At that point the epidemic could have been stopped if urgent action had been taken. The WHO said that it was all under control, and that the outbreak would stop. How wrong that prediction was.

A team of scientists, led by Robert Garry from Tulane University, had been undertaking research at a hospital in Sierra Leone. The team reported that the number of Ebola cases was being underreported, and that many people sick with the virus were dying before they could reach a hospital. By August, five of the team of fifty-eight scientists had died of Ebola.

By 10 April, international aid organisations were implementing emergency measures across West Africa to try to stop the spread of the epidemic. Ebola had already spread to Conakry, the capital of Guinea, and to Liberia. In May, Sierra Leone confirmed the first deaths from Ebola and began to restrict travel to areas that were badly affected. In June, the WHO confirmed that the Ebola outbreak was the deadliest ever with a 60% increase in cases within two weeks. By June 23rd, the outbreak was declared out of control, and Medicins sans Frontieres had identified sixty hot spots of Ebola infection.

The media in the West finally woke up to the threat of Ebola when, on July 30th an American citizen collapsed and died in Nigeria. Patrick Sawyer worked in the

Liberian Ministry of Finance, but was a naturalised American citizen with a wife and two children in Minnesota. He had been caring for his sister, who was ill in Liberia, not realising that she had Ebola. Patrick Sawyer's death mobilised Western governments into acting to help stop the epidemic.

Aid workers were becoming infected with Ebola, and there was much concern in Western countries that people returning from West Africa could take Ebola virus home with them. Nigerian authorities were attempting to trace all the people that Patrick Sawyer had been in contact with, a stunning 30,000. There was concern that any of the passengers Sawyer had travelled with could be infected with the virus.

Dr Joen Breman said that, "An African epidemic might be compared to a giant tree falling in the forest. Nobody notices it has fallen. When the first white person dies, the epidemic begins", When Western governments began to realise that Ebola was only an aircraft's flight away from their own countries did any positive response begin.

Across the world countries announced stringent measures to control the outbreak, and to prevent Ebola spreading. Sierra Leone declared a state of emergency on July 31st, and the WHO stated that the death toll from Ebola had risen to over 700.

In August, the head of the WHO, Margaret Chan and the Presidents of the West African countries involved in the outbreak met to launch a $100 million response. Hundreds of medical personnel were deployed to help over-stretched health workers on the front line. Margaret Chan said that Ebola was moving faster than efforts to control it. By the end of July, there were 1323 confirmed cases of Ebola.

The media around the world began to publish articles and guides on how to protect people from Ebola. It was stressed that it was only by being in contact with the bodily fluids of Ebola sufferers that the virus would be passed. This did not prevent hysteria in many places. A doctor at the Korle Bu Teaching hospital in Ghana said that, "I think that many doctors will run away at the emergency if a patient with suspected Ebola comes in. I mean we really, really don't feel safe". Newspapers declared that, "Ebola is just one plane ride away from western countries", and showed flight times from West Africa to Europe, with routes. Virgin and British Airways stated that they would continue flights into and out of the affected areas, but that the situation was being carefully monitored.

Liberia declared a state of emergency on 7th August that would last for 90 days. It would allow the government to curtail citizens' civil rights by imposing quarantines on affected areas. In Sierra Leon, police and soldiers blockaded roads in ands out of rural areas to prevent the spread of Ebola.

On August 14th, it was reported that Guinea had declared a state of emergency, This meant tighter border controls; the isolation of anybody suspected of having the Ebola virus; and a ban on moving bodies from village to village. It is the custom to bury a dead person in their home village. The risk of moving a dead body, loaded with Ebola virus, from one village to another was too dangerous.

On 11th August there had been 1975 confirmed cases of Ebola in Guinea, Sierra Leone, Liberia and Nigeria, with 1069 deaths. By 19th August, the numbers had risen to 2240 cases, with 1229 deaths.

There were reported cases of Ebola in Europe, Asia and North America: none of these cases turned out to be the virus. In the UK, the Independent newspaper reported that Ann Coulter., a conservative in the US, had stated that a US missionary who went to help in Africa was, "idiotic and narcissistic" for helping. Donald Trump echoed her views, "The US cannot allow Ebola infected people back. People that go to far away places to help out are great, but must suffer the consequences", he Tweeted. These comments were in response to the news that two US aid workers were being flown back to the US for treatment as they had become infected with Ebola while caring for sick people in Africa.

ONE.org (whose purpose is to fight extreme poverty and preventable disease around the world), asked people to support the humanitarian response to Ebola, and listed just some of the organisations already delivering vital services in West Africa:

Action Aid: Working in Sierra Leone to inform people about how to protect themselves and their families.

Caritas: In Guinea delivering soap, chlorine and other materials; teaching people about good hygiene. Soap and chlorine, which are effective at killing the virus, has been distributed to more than

100,000 people.

International Federation of Red Cross and Red Crescent Societies: In Sierra Leone, red Cross teaches people how to protect themselves: many are in denial about Ebola.

Medicins sans Frontieres: Have 676 staff working in Guinea, Sierra Leone and Liberia working to save lives.

Plan: Running public health campaigns in Guinea, Sierra leone and Liberia. Setting up hand-washing stations at schools, health posts and other public facilities to keep people safe.

UNICEF: Working in Guinea, Sierra Leone, Liberia, Guinea-Bissau, Senegal, Mali and Gambia to prevent the spread of Ebola with TV and print campaigns. Over 600,000 bottles of chlorine and 2 million bars of soap have been distributed since April.
The WHO emergency committee held two days of talks in Geneva with heads of state from the affected countries, and declared a public health emergency. It was recommended that all travellers leaving the affected countries be screened for fever, and that no corpses should be transported across borders. The charity Save he

Children said that it was scaling up operations in the region, and that their resources were already being overwhelmed. Orphaned children were being ostracised by their communities and needed urgent help. The WHO stated that anybody visiting the affected areas, or who had been in contact with a possibly infected person, should know that the transmission of Ebola requires direct contact with blood, bodily secretions, organs or other body fluids of infected living or dead persons or animals. The US was sending teams of experts to Liberia, including twelve specialists from the CDC.

Scientists observed that outbreaks of tropical diseases often occur in the rainy season in west Africa. Liberia is one of the wettest countries in terms of annual rainfall. In the villages, there is no running water, so no flushing toilets. People share latrines that are pits dug away from the village; when one is full, it is covered over, and another is dug. Hand washing is difficult - in villages where wells have been provided by some of the charities like Water Aid, families can take water back to their homes for washing and cooking, but without a well, river water is used, often dirty and full of bacteria. In the shantytowns there are open sewers that are breeding grounds for diseases.

Health care is difficult as many of the villages are isolated, and a considerable distance from a hospital. Missions often have a clinic, but often missionaries have only basic training in health care, and cannot cope with epidemics, especially on the scale of the current Ebola outbreak. Civil wars in the region resulted in a loss of doctors and nurses: in 2010 there were only 51 doctors in the whole of Liberia.

By 6th September, the deaths from Ebola had passed 2000, with the number of people infected expected to be 20,000 by the end of the year. The healthcare workers were doing their best to cope, but many had died, and nobody seemed to be able to stop the spread of the virus. The BBC condemned the response to the outbreak as "lethally inadequate", while the New York Times called for the United National Security Council to pass a resolution giving the UN total responsibility for controlling the outbreak. Countries were talking about sending beds and medical supplies to the region, but without an organisation to coordinate the response, nothing much would happen.

The government in Sierra Leone decided on

quarantine measures on 25th September. More than one million people were quarantined in an effort to stop the spread of the Ebola virus. The General secretary of the United Nations, Ban ki-moon, finally convened a meeting at the UN for world leaders to discuss what should be done. President Barack Obama had already decided to send 3000 troops to west Africa to help health workers. They would build hospitals and set up logistics. The CDC predicted that there could be 1.4 million cases of Ebola in Liberia and Sierra Leone by January 2015.

A summit conference was held in London at the beginning of October to discuss the global response to the Ebola crisis: by then, 3338 people had died, and there were 7178 confirmed cases. The charity Save the Children, based in the UK, pledged £43 million, with £3 million earmarked for Sierra Leone. Comic Relief in the UK pledged £1 million. 160 National Health service staff were travelling to Sierra Leone answering a call for volunteer health workers.

In the affected countries, Myth Busting messages were broadcast in the media:

Eating raw onions will not stop you

catching Ebola
Drinking salt water will not stop you catching Ebola
Drinking condensed milk will not stop you catching Ebola
You can get Ebola from a dead person
You can get Ebola from having sex
Ebola is present in semen
Ebola can be present for 7 weeks in semen after having Ebola
Health care workers do not carry Ebola
Keep washing your hands with soap and clean water

The London conference determined that there were five ways to stop the Ebola epidemic:

- More treatment centres
- Home care. Teach communities hoe to care for people with Ebola
- Air bridge and medevac system needed. Supplies are desperately needed in affected countries
- Preparation elsewhere - countries in the region must be ready to deal with cases of Ebola. Public awareness programmes and medical kits to be available.
- Vaccines are desperately needed.

It was determined that the next countries most at risk from Ebola are Mali, Guinea-

Bissau, Ivory Coast (Cote d'Ivoire), and Senegal.

Western countries began to screen air passengers by checking temperatures. The US already ha this in place, but the UK was still preparing questionnaires for passengers.

There were chaotic scenes in Spain where a nurse who had cared for two Spanish missionaries who died of the Ebola virus had become ill. Health workers at the hospital where the events took place, said that they had been provided with substandard protective clothing, and that training for doctors, nurses and ambulance workers in how to cope with patients with Ebola was inadequate.

In the US, a passenger on a four-hour flight from Philadelphia to the Dominican Republic, began coughing and sneezing, and shouted that he had Ebola. There was panic among the passengers, and officers in full biohazard suits and masks removed the man from the aircraft when it landed. The remaining passengers were all checked as a precaution. The man who started he panic was described as 'unbalanced'. Two other flights, one in Australia, and another in the US, both had to make emergency landings

with passengers who were feared to have symptoms of Ebola. A woman on a coach in the UK became feverish after boarding the coach. The National express bus station in Liverpool was closed, and paramedics in full body suits and protective masks cared for the woman.
The other passengers fled in terror. In the US, people took to wearing biohazard suits on flights to protect themselves against Ebola.

Meanwhile in Texas, Thomas Eric Duncan was visiting Dallas from Liberia to see his sister. He seemed well, but four days after arriving in Dallas, he was burning with fever, had vomiting and diarrhoea - classic symptoms of Ebola. He began to feel unwell on September 25th, and went to the Emergency room at texas Health Presbyterian hospital. The nurse who dealt with him did not respond to the information that he had recently arrived from Liberia, and sent Mr Duncan home. On September 27th, he was so ill that he had to be taken to the same hospital by ambulance. It took two days to confirm that his illness was Ebola, and the CDC sent a team to Dallas to locate every person who had been in contact with Mr Duncan, and exposed to the virus. 18 people were found, including school children who were neighbours in the apartments where his sister lived. Despite being treated with an experimental vaccine, Mr Duncan died on 10th October.

In the UK, a British nurse, William Pooley, had been airlifted from Sierra Leone after being infected with Ebola.
He was taken to the specialist isolation unit at the Royal Free Hospital in London. The experimental vaccine, ZMapp was used to treat Mr Pooley, and he responded well, and was able to get back to eating bacon butties, as reported in the press.

In the US, a nurse, Nina Pham, who was one of the nurses who had ben caring for Thomas Duncan, contracted the virus. It is thought that Ms Pham may have caught the virus when she removed the bulky protective gear that health workers have to wear to treat patients with Ebola. The protective gear has to be removed without any of the material touching the skin, eyes, mouth or nose. A survey of 1900 nurses by the National Nurses United union in the US, found that an astonishing 76% had not received any training, or policies relating to patients presenting at their hospitals with the symptoms of Ebola.

In just 13 days, from 2nd October to 15th October, the death toll from Ebola had risen to 4450, with more than 8900 people infected. The WHO stated that the true figure was probably much higher as many people die before they can reach a hospital,and some have been hidden by relatives who feared that going in to a hospital would be a death sentence for their loved ones.

The WHO announced on 14th October that Ebola cases could number 10,000 a week by December. The death rate had risen to 70% for people who contracted the virus, and there seemed to be no way to halt the inexorable rise of Ebola cases.

The head of Medicins sans Frontieres, Sharon Ekambaram, asked, "Where is WHO Africa? Where is the African Union? We've all heard their promises in the media, but have seen very little on the ground". The WHO stated that the target was to isolate 70% of cases in order to treat them over the next two months. Mark Zuckerberg, the founder of Facebook, announced that he and his wife, Priscilla Chan, were donating $25 million to the CDC to help in the fight against Ebola.

The CDC then announced that it had staffed and equipped extra testing laboratories in the stricken countries:

Liberia had 5; Sierra Leone had 4, and Guinea 3. The bad news was that the laboratories could only deal with 100 samples a day. Faster ways to diagnose Ebola were desperately needed. If a rapid test for Ebola could b found, it would prevent people who were sick with another disease, being placed in hospitals with Ebola patients, making it more likely that they would also become infected.

In the US in the best laboratories, it can still take 8 hours to process a blood sample; in West Africa it can take days. Taking blood for testing is also a risky business - anything involving the use of needles is dangerous. Around a disease like Ebola. Dr Estella Lasry, a doctor with Medicins sans Frontieres said, "Taking samples is extremely dangerous. Sometimes the patients are cooperative, but sometimes they are combative. At any time you risk a needlestick injury that can expose you to the virus". Samples are needed because Ebola and malaria look very similar in the early stages: both have symptoms of high fever and pain.

The US company Corgenix had received a $2.9 million grant in June from the national Institutes of Health to develop a pinprick test for Ebola. A pinprick is all that would be needed, and it would reveal a dark red line on a test paper within 15 minutes. The test could be used all over the world: especially at airports where a simple pinprick test could determine which passengers were positive for Ebola.

The experimental Ebola drug ZMapp had been given to two American aid workers who were infected: they both survived. Production of ZMapp was being increased, but it would take time to be ready to use on Ebola patients. Two more vaccines for Ebola were being tested on volunteers in the US, Mail and Switzerland.

The village of Gbantama in Sierra Leone decided to take matters into their own hands, and decided to build their own Ebola treatment centre. The Paramount Chief of Gbantama, Baiburch Sallu Lugbu, said that talk of the sickness being caused by witchcraft had stopped because people are caring for their own, and are not watching sick relatives being taken away to die in a strange place. Aid organisations supported the village, and another 20 similar facilities are being planned. Many were still dying, but people were being trained how to take care of others without catching the virus themselves.

The Medicins sans Frontieres Ebola crisis update, issued on 21st November, showed that the number of deaths had risen to 5420, with 15145 cases. The virus was now in Senegal and mali. The good news was that three different treatments were to be tried. The French National Institute of |Health and medical research (INSERM), would begin a trial of a new antiviral drug Favipiravir at Medicins sans Frontieres' facility at Gueckedou, Guinea. The Antwerp Institute of Tropical Medicine (ITM), would undertake a trial of 'convalescent whole blood and plasma therapy' in Conakry, Guinea. The University of Oxford would lead a trial in behalf of the International Severe Acute respiratory and emerging Infection Consortium (ISARIC), of the antiviral drug Brincidofovir; a site had not yet been decided on.

The UN Mission for Ebola Emergency Response (NMEER), was to be based in Ghana with five strategic priorities:

- Stop the spread of Ebola
- Treat infected people
- Ensure essential services were in place
- Preserve stability in affected countries
- Prevent the spread of Ebola to uninfected countries

The spread of Ebola in Monrovia, Liberia, showed signs of slowing down as only 18 cases had been confirmed in the past week. Treatments for malaria had been stepped up, and 100,000 households had received medication. The second round of treatment would start
In late November.

Unfortunately, the spread of Ebola was continuing in Sierra leone. The UK and China had sent teams to construct new treatment centres around the country. In Freetown, it was planned to distribute anti-malarial medication to 1.4 million people, with a second round a month later.

There was good news from Nigeria, and Senegal, who reported that there had been 42 days without a case of Ebola being reported. The governments of Nigeria and Senegal would remain vigilant.

After the encouraging news that international governments and the WHO were finally sending aid and support to Ebola-stricken countries in West Africa, and that the number of cases appeared to be falling, there was discouraging news from Sierra Leone. A new outbreak of Ebola was reported in the remote area of Kono that borders Guinea . At least 87 people had died, and there were hundreds of new cases. Scientists believed that it would be the middle of 2015 before Ebola was under control.

British health workers at the Moyamba Centre, built by Royal Engineers in eastern Sierra Leone, objected to new protocols put in place for when they return home. There were 780 NHS volunteers in the region, and they faced increased restrictions on their movements on returning to the UK. Health workers from charities are not allowed to return to work for three weeks after they return home, and Public Health England was bringing NHS practices into line with them.

Within nine days, a Scottish nurse was being transferred to the Royal Free Hospital in London from Glasgow after developing Ebola on her return home from Sierra Leone. Health officials were trying to trace all the people that the nurse had been in contact with on her flights home. The nurse had been cleared at the screening process at Heathrow airport, but developed symptoms of Ebola when she returned home to Glasgow. There was an immediate outcry in the press, with a doctor who had been sitting next to the nurse on the flight from

Sierra Leone to the UK, stating that the screening process was 'totally inadequate'. Dr Martin Deahl also criticised Public Health England's guidelines for returning health workers as 'utterly illogical'. PHE confirmed that it would be reviewing its procedures and advice.

Nurse Pauline Cafferkey had told officials at Heathrow airport, London, that she had a fever. They took her temperature seven times, but then allowed her to continue her journey home to Glasgow. Nurse Cafferkey was treated with plasma from Ebola survivor and fellow nurse William Pooley. Nurse Pauline Cafferkey left hospital on 23rd February, free of the virus, only three weeks after being in a critical condition. She said that the Royal Free hospital has saved her life. In the US, returning health workers face far stricter control measures that include a ban on using public transport, and 21 days in quarantine at home.

On the last day of 2014, the WHO reported that the number of people infected with Ebola had risen to over 20,000, with more than 7900 deaths.

Vaccines and possible treatments

The WHO declared in August 2014, that it was ethical to use experimental drugs to try to halt the spread of the Ebola outbreak. It has been known for some time that convalescent serum had been used on some Ebola victims, using blood products from donors who had survived infection with Ebola. During the last outbreak in 2012 in the Republic of Congo, eight people were given convalescent serum and seven survived, so it is worth trying to save people with this method.

There are several experimental vaccines:

- ZMapp - is the result of a collaboration between Mapp Biopharmaceutical, the Public Health Agency of Canada, Defyrus, LeafBio of San Diego, Kentucky BioProcessing, and the US Army Medical Research Institute of Infectious Diseases (USAMRIID). The drug is composed of three monoclonal antibodies manufactured from the nicotine plant, Nicotiana. The antibodies attack proteins on the surface of the Ebola virus.

- TKM-Ebola - developed by Tekmira Pharmaceuticals in Canada. The drug targets strands of genetic material in the virus.

- An RNA drug similar to TKM-Ebola that has been developed by Sarepton Therapeutics.

Nicotiana rustica was the original tobacco plant brought back to the Old World from the New World by Walter Raleigh. It contains very high levels of nicotine, and is quite toxic. It was superceded by the milder Nicotiana tabacum, that is less toxic to smoke.

ZMapp is produced from harvested tobacco plants; antibodies that have been produced in the plants have to be purified and prepared in a liquid state for injections. In 2014, ZMapp was shown to be effective in animal trials, and production is being accelerated to produce enough product to vaccinate as many people as possible.

Some in the US criticised the use of ZMapp on people until it had undergone stringent testing procedures, but with the number of cases of Ebola in West Africa rising again, there may not be time to undertake all the usual tests which can take years.

In November 2014, Medecins sans

Frontieres announced that clinical trials would be carried out in three of its treatment centres in West Africa. The French National Institute of Health and Medical Research (INSERM), would trial the antiviral drug Favipiravir in Gueckedou, Guinea; the Antwerp Institute of Tropical Medicine (ITM) would trial convalescent whole blood and plasma therapy at the Donka Ebola Centre in Conakry, Guinea; and the University of Oxford would trial, on behalf of the International Severe Acute Respiratory and Emerging Infection Consortium (ISARIC) and funded by the Wellcome Trust, the antiviral drug Brincidofovir at a site that was not yet determined.

- Favipirovir was licensed in Japan to treat influenza. There have been promising tests on rodents.

- Brincidofovir has been developed for use on the DNA of viruses like smallpox and herpes. There is limited data on Ebola.

A third antiviral drug, Lamivudine, has been used against Hepatitus B and HIV/AIDS. Will it work against Ebola? Favipirovir and Lamiduvine were developed for use on small RNA viruses, similar to the Ebola virus, so will hopefully be effective. Brincidofovir was used to treat Thomas Duncan in Texas. Unfortunately, he died, but if used during the early stages of Ebola it may be more effective.

Three groups of 9,000 people (27,000) will be given the vaccine in Liberia, as it is the worst affected country.

The BBC announced on 7 January 2015 that various trials were underway in West Africa. Brincidofovir was to be assessed on a voluntary basis in Liberia; and Favipirovir in Guinea. Oxford University and Tekmira Pharmaceuticals Corporation from Canada, were to test the new TKM-Ebola-Guinea therapeutic in Guinea, led by the Antwerp ITM. A British nurse, Pauline Cafferkey, infected with Ebola in West Africa, was receiving the drug at the Royal Free Hospital in London.

The University of New South Wales, Australia has championed the use of Lamiduvine, citing the use of the drug by Dr Gorbee Logan in Liberia; fifteen of his patients were treated with the drug, of whom thirteen recovered from Ebola. As UNSW stated, "In a situation as dire as the one currently occurring in West Africa....

any intervention that saves lives is very much welcome". The UNSW points out that this effectively cuts the fatality rate from 70% to just 13%. Why, therefore, are the WHO and the CDC not paying attention to Dr Logan's results? Is it because Dr Logan is rural African doctor? Nothing further has been heard about his breakthrough. Lamiduvine is not on the list of Ebola vaccines and therapies, published by the WHO on 21 January 2015.

The Tekmira Pharmaceuticals Corporation published a periodic update on its TKM-Ebola programme on 21 October 2014. The update states that Tekmira was actively engaged with an international consortium led by ISARIC, the University of Oxford, the WHO, the CDC, MSF, the Institute Pasteur, and others, to undertake clinical trials of TKM-Ebola in West Africa. The Wellcome Trust had awarded £3.2 million to the Consortium to undertake the work. At the bottom of the page of the update, there is a paragraph headed, 'About Joint Project Manager Medical Countermeasure Systems (JPM-MCS).'

It states that the work is being funded by a $140 million contract with the US Department of Defense, JPM-MCS. It is a component of the Joint Program Executive

Office for Chemical and Biological Defense. This is with the aim of providing US military forces with safe and effective solutions to - 'counter chemical, biological, radiological and nuclear threats'. The funding for anti-biological warfare countermeasures is $140 million, while £3.2 million is awarded to help fight Ebola.

Tekmira also announced, in December 2014, that a new product, TKM-Ebola-Guinea, had been designed to exactly match the gene sequence of Ebola Guinea, with RNA interference. Tekmira had originally been working with Ebola Zaire from the Kikwit outbreak in 1995, and was modified to counter the Ebola Guinea strain. Experiments with non-human primates demonstrated that when the new therapeutic product was used, it was completely successful: "100% protection an otherwise lethal dose of Ebola Zaire virus."

The results were published in 2010, so why wasn't the product being used during the 2014 outbreak?

The article concludes, "In March 2014, Tekmira was granted a capital Fast Track designation from the US Food and Drug Administration for the development of TKM-Ebola".

After 9,700 deaths from Ebola in west Africa, why wasn't TKM-Ebola-Guinea used?

Conspiracy theories or the truth about the Ebola outbreak?

a) Why is NASA interested Ebola ?

There have been a number of papers on the Ebola outbreak, written at the NASA Goddard Space Flight Centre. The first paper examines satellite detection of vegetation fluctuations that appear to be associated with outbreaks of Ebola. Areas that have been identified from measurements made by the Advanced Very High Resolution Radiometer carried on meteorological satellites, show that Ebola appears to coincide with the rainy season in West Africa.

A second paper stated that Ebola appeared following 'sudden transitions between dry and wet seasons'. The findings were supported by data from two of the three outbreak sites in 1994, but that further studies were needed.

But, why were NASA scientists interested in Ebola outbreaks? One writer suggests that NASA's interest is connected to the manufacture od space suits. Scientists examining the Ebola virus have to be securely protected from contamination, as do the health workers. The same concerns can be applied to handling extra-terrestrial samples that may contain 'evidence of life', microbiological or otherwise'. Health officials dealing with the Ebola virus have been considering 'the incorporation of robots and advanced protective suits'.

It was also suggested that the CDC's Ebola procedures could be used in the investigation
of samples from Mars or other missions that NASA undertakes. A sample of soil from Mars would have to be securely contained before, during, and after the voyage back to Earth. Nobody knows if soils from Mars contain evidence of microbiological existence. On the other hand, the protocols developed by the CDC for the Ebola outbreak will also be used to ensure that extraterrestrial environments are not contaminated by microbes from Earth.

The space suit, and microbe handling theories are interesting, but do they really answer the question of why NASA has been spending a lot of dollars to look at vegetation from space.

b) Theories about the 2014 Ebola outbreak

The suggestion that viral outbreaks could have been started accidentally by military bioweapon experiments is not new. President Richard Nixon made a speech to Congress in July 1969, on the problems caused by population growth. Nixon specifically mentioned countries in Latin America, Asia and Africa, and stated that, "In sum, population growth is a world problem which no country can ignore, whether it is moved by the narrowest perception of national self-interest or the widest vision of a common humanity". It has been suggested by some, that covert military operations followed this speech in an effort to curb population growth.

Leonard Horowitz states that in 1976, there had been covert military operations that had permeated central Africa at the time of the major Ebola outbreaks. There were numerous USAID-sponsored programmes and widespread vaccine experiments throughout many African countries. Horowitz also suggests that biological weapons contractors were paid millions of dollars during the 1960s and early 1970s, to produce and test 'system-depleting' viruses. These would include the viruses with the pathological effects of the haemorrhagic fever viruses Ebola and Marburg. He also refers to the HIV-AIDS virus being man-made. The primates first infected with the viruses could have been part of a covert operation.

The outbreak in Gabon in 1996, began around the same time that, 'some French soldiers came up to Mayibout 2 in a Zodiac craft and camped near the village'. Some hunters from the village found a dead chimpanzee, butchered it, and shared the meat in the village. After that anybody who had eaten the chimpanzee became very ill. The surviving villagers suspected that the French soldiers had used a biochemical weapon to kill the chimpanzee, Was this just a wild theory to explain how villagers had become infected with Ebola, or ws there some truth in the story? It is known that the French were researching biological weapons, including anthrax, botulism toxin, cholera, rinderpest, and salmonella. In the 1960s, French scientists were manufacturing Sarin and XV nerve agents. The French destroyed their stockpiles of biological weapons after the 1972 Convention on the Prohibition of the Development, Production and Stockpiling of Bacteriological (Biological) and Toxin Weapons and on their Destruction was signed on April 10th, 1972. Can the public trust any country that states that it has destroyed all its stockpiles? There is always the threat that the enemy will hide some of their unpleasant weapons so that UN inspectors cannot find them. That was the premise for the invasion of Iraq - that Saddam Hussein had stockpiled 'weapons of mass destruction'. It is also easy to remember the Greenpeace ship 'Rainbow Warrior', blown up in New Zealand by French secret agents.

Jennifer Gadarowski has stated that many countries have Biolevel 4 laboratories: necessary to work with dangerous, infectious agents that have no vaccines. They are: Bolivian and Argentinian haemorrhagic fevers; Marburg virus; Ebola virus; Lassa fever; Crimean-Congo haemorrhagic fevers, and smallpox; the last named may have been eradicated as a scourge for humankind, but it is still very much alive in a number of laboratories.

Bio-agents are studied for defensive purposes. In 2008 countries involved in such studies were China, Cuba, Egypt, Iran, Israel, North Korea, Taiwan and Syria. With the rise of Islamic state in the Middle east, it is to be hoped that IS never locate any of Syria's Level laboratories. The commander of a Syrian rebel group took a laptop from an Islamic State hideout. The IS fighters had run away and left the laptop behind. The laptop contained manuals on how to make bombs; how to steal cars; how to use disguises, and most worrying of all, how to develop biological weapons, including bubonic plague. Also included was a document with a fatwa, "If Muslims cannot defeat the kafir (unbelievers) in a different way, it is permissible to use weapons of mass destruction,. Even if it kills all of them and wipes them and their descendants off the face of the Earth". The fatwa was issued by Saudi cleric Nasir al-Fahd, at present in prison in Saudi Arabia. The document also advises putting biological weapons like plague and Ebola, inside grenades and throwing them into places where people gather, like football stadiums, or next to air conditioners.

The Grand Mufti of Egypt, Ali Gomaa, has also issued a fatwa against the use of weapons of mass destruction. It stated that, "This constitutes a surprise attack, and killing of the unaware. It is not sanctioned to kill them. This act would necessitate killing and annihilating Muslims in those countries, which is unlawful in Islam". The killing of fellow Muslims does not seem to bother the followers of Islamic State, who have killed their fellow believers in Syria and Iraq.

Vladimir Nikiforov, head of the Department of Infectious Diseases at Russia's Federal Medical-Biological Agency, said that it was comparatively easy to use Ebola virus as a weapon in a spray. Dr Peter Walsh, at Cambridge University, agreed with Dr Nikiforov, stating that there are just a few laboratories in the world that are capable of handling something as dangerous as the Ebola virus, and they are extremely well protected. The risk may be that a group like IS manage to obtain the virus from West Africa.

On 2nd January 2015, there were reports that IS fighters had been infected with Ebola, and had gone to a hospital in Mosul, Iraq. The infected men were Africans, who had travelled to join IS before becoming ill. IS have killed more than a dozen doctors in Mosul, so confirmation that the disease was Ebola would be difficult. There were fears that IS would send fighters infected with Ebola to the UK aas 'human bug bombs'. However, a later report from Breikbart stated that IS had incinerated five fighters who had Ebola to prevent the spread of the disease: whether they were alive or dead before being burned was not reported.

The US Army Medical Research Institute of Infectious Diseases (USAMRIID) is based at Fort Detrick, Maryland. According to its website (accessed 4 February 2015): 'USAMRIID's mission is to protect the war fighter from biological threats and to be prepared to investigate disease outbreaks or threats to public health. A team from USAMRIID have been working in Sierra Leone and Liberia since 2006, investigating Ebola evidence. Samples from suspected Lassa fever cases showed that two-thirds of the patients had also been exposed to dengue fever, West Nile virus, Yellow fever, Rift Valley

fever, Chikungunya, Ebola and Marburg virus. Those that were positive for Ebola were for the Zaire strain, which was completely unexpected. Randal J. Schoepp, Applied Diagnostics Branch Chief Diagnostic Systems Division, USAMRIID, helped to set up a diagnostic laboratory in Sierra Leone; he said that only one case of Ebola had been reported previously in the region, and that was the Tai Forest Strain, and not Ebola Zaire virus. The scientists from USAMRIID were investigating emerging viruses, but were also assisting countries in west Africa by building diagnostic laboratories to help in the fight against emerging viruses.

The presence of the US military and of USAMRIID scientists in West Africa can be explained by the work that they have been doing. They are not working alone: Tulane University scientists have also been carrying out research, developing diagnostic kits for Lassa fever, for which the University received a $7,073,538 grant in 2009 from thhe National Institute of Health. The same organisation donated $2.9 million to Corgenix in June 2014 to develop a pinprick test for Ebola.

On the other hand, there have been stories and articles on the Internet about a US Level 2 bioweapons research laboratory inside the hospital at Kenema, Sierra Leone. This is the same laboratory that Randal J. Schoepp helped to establish. The laboratory is supported by the US Armed Forces Health Surveillance Center, and there are rumours that research undertaken there led to an accident in which Ebola virus escaped and began to infect the surrounding area. Reuters reported that an angry crowd had gathered outside the hospital in July 2014, and threw stones at the windows. The crowd had to be dispersed with tear gas, and a nine-year-old boy was shot in the leg by the police. Reuters did not state that the crowd was demonstrating against leaked viruses, but about rumours of 'cannibalistic practices' on the wards, reported by a nurse: a completely different thing. What the practices were alleged to be, the report does not say.

There are rumours and articles about bioweapons being made using Ebola virus, but factual information is available, with details about the work being undertaken by the laboratory. A Facebook page, dated 8 March 2011, refers to the Lassa fever Programme at Kenema Government Hospital, being funded by the Viral Haemorrhagic Diseases Consortium. A journal article entitled, 'Undiagnosed acute viral febrile illnesses', in the Emerging Infectious Diseases, in July 2014. The article was co-authored by, among others, Randal J. Schoepp and Sheik H. Khan. Sadly, Dr Khan died of Ebola at the end of July.

There are also stories around the shooting down of a Boeing 777 Malaysia Airlines aircraft over Ukraine in July 2014. One of the passengers was Glen Thomas, a consultant in Geneva, who was an 'expert on AIDS and the Ebola virus'. Mr Thomas was on his way to Melbourne, Australia, to take part in a conference. Was he about to make an important announcement: that the laboratory had manipulated results to show that more people were infected with Ebola than there actually were. This was so that the local population could be used in an experiment to test vaccines for Ebola. We shall never know what Glen Thomas was going to say at the conference. It seems extreme to shoot down an aircraft in order to kill one man, especially over a war zone. Or was it over a war zone so that other people could be blamed for the disaster? This is how conspiracy theories begin!

c) Why does the Centers for Disease Control own a patent on Ebola?

Why does the CDC own a patent on Ebola? The CDC own patent no: CA2741523A1, dated 2010. It is for 'EbolaBun', which is Ebola Bundibugyo. The patent is described as, "The present invention is based upon the isolation and identification of a new Ebola virus species, EboBun. EboBun was isolated from the patients suffering from haemorrhagic fever in a recent outbreak in Uganda".

Ebola Bundibugyo first appeared in 2007, in Uganda. A mysterious illness had broken out in Bundibugyo, and people were dying. Blood samples were sent to the CDC, who found that it was a new Ebola virus, that was at least 32% different from the other viruses in the Ebola family. A team of scientists went to the area to investigate the new ebolavirus. By the time the outbreak ended in January 2008, 149 people had been infected, of whom 37 had died, a fatality rate of 40%. The Bundibugyo Ebola virus kills more older people (aged 20-70) than the Sudan and Zaire strains, but it is not as lethal.

Why did the CDC decide to patent the Bundibugyo Ebolavirus? Some have suggested that it is because the US government and private companies intend to engineer many different strains of Ebola in order to create vaccines that can then be sold to countries, and other governments, at vast profit.

It has also been suggested that it is one of the reasons that Americans suffering from Ebola are transported back to the US for treatment. Adams argued that, "These patients are carrying valuable intellectual property assets in the form of Ebola variants, and the Centers for Disease Control clearly desire to expand its patent portfolio by harvesting, studying and potentially patenting new strains or variants". Some scientists have predicted that Ebola could become an airborne contagion. If that happens, the numbers of infected people could number into the millions, as in the influenza outbreak of 1918. Drug companies who marketed a vaccine would make billions in profits.

Adams reported that shares in Tekmira, a Canadian pharmaceutical company, surged over 11% in one day as news that the US Federal Drugs Agency was being pressured to fast-track ab Ebola vaccine trial. Tekmira announced, in October 2014, that the company had begun limited manufacture of a 'therapeutic' for the Ebola Guinea variant. The company's scientists have developed a method for silencing disease-casuing genes. Tekmira is part of an international consortium, led by the International Severe Acute Respiratory and Emerging Infections Consortium (ISARIC), The University of Oxford, the WHO, the CDC, Medicins sans Frontieres, Institut Pasteur, and others. The Wellcome Trust awarded £3.2 million to the Consortium to fund the initiative.

The Institut Pasteur identified a new variant of Ebola virus in Guinea. The Ebola outbreak is of the Zaire ebolavirus, but it is a slightly different variant to the previous outbreaks in Republic of Congo and Gabon. Blood samples taken from patients in Guinea showed that the Ebola virus they are suffering from was 97% identical to the strains already seen in Republic of Congo in 1976 and 2007, and in Gabon in 1994 and 1996. The present outbreak has been caused by a new variant of Ebola in Guinea, that has now been called Ebola-Guinea.

The silence of leading health organisations on matters related to the ongoing outbreak of Ebola in West Africa is puzzling.. In September 2014, the University of Minnesota suggested that Ebola could mutate to become an airborne virus. It is already known that Ebola Reston can be transmitted by airborne particles, and there is no reason to suppose that Ebola Zaire could not mutate in the same manner. The CDC admitted in October 2014, that droplets from a sneeze could spread Ebola. The UK's Defence, Science and Technology Laboratory (DSTL), discovered that droplets of Ebola Zaire virus were still alive 50 days later, although they were kept at low temperatures. The CDC stated that the virus could live for a few hours in puddles of water, or bodily fluids. The CDC then did an about-turn, and said that Ebola could not be transmitted through droplets, presumably to prevent panic if an airline passenger started sneezing, and looked very ill, for example. The National Institute of health Director, Dr Anthony Fauci, said that while viruses mutate all the time, it is very rare for a virus to change its mode of transmission. He used the example of HIV, which has killed around 35 million people, and infected 77 million, and it has never changed how it is transmitted. It is to be hoped that the Ebola virus will be the same.

Legal matters related to Ebola

One of the most bizarre reports to emerge from the 2014 Ebola outbreak, was the news that a doctor was threatening legal proceedings against Teresa Romero, a Spanish nurse who contracted the disease while tending a Catholic missionary in Madrid who had been infected in Liberia. The doctor alleges that Ms Romero did not tell him that she had been working on a ward where Ebola patients were being treated.. The GP prescribed paracetamol for her fever, not realising that she was showing the symptoms of infection with Ebola.

Teresa Romero has sued the Health Council of Madrid. She says that she did not receive enough training on how to protect herself against Ebola while treating patients. A senior official of the Health Council allegedly said, "You don't need a Master's degree to learn how to put a suit on". Ms Romero is also seeking €150,000 for the loss of her pet dog, Excalibur, who was euthanased by the authorities in case he also had Ebola.

The Ebola outbreak is also causing problems for airlines, shipping companies and international trade. Airlines and shipping companies have to be aware of legal issues around risks to passengers and crew from Ebola. Some countries have been declared unsafe for airline passengers and crews, and flights have been abandoned until the Ebola outbreak is over. The CDC gave guidance to airlines to stop passengers from boarding aircraft if they had been in Ebola affected countries, and especially if they had any symptoms such as fever, severe headache, muscle pain, stomach pain, or unexplained bruising. If travellers became ill during a flight, the crew should follow infection control precautions like keeping the passenger separated as much as possible from the rest of the passengers, and wearing a mask and gloves with protective gown or apron.

For ships the situation is a little different as generally voyages between ports take longer than air flights. The captain of a ship has a duty of care to the crew and passengers, and it can be particularly hazardous if workers loading or unloading a cargo become ill in a country with cases of Ebola, or if a stowaway is discovered who is sick.

There have been reported cases of ships' crews refusing to enter ports where an infectious disease has been reported. The first UK Quarantine Act was passed in 1712 mandating a 40 day period before a ship could enter a port that had epidemic disease. In the past few years , there have been outbreaks of Norovirus and Legionella on cruise ships. Most countries have laws in place to detain ships in circumstances where the health of those on board, or those on land, is at risk.

The ports of |Conakry (Guinea), Freetown (Sierra Leone), and Monrovia (Liberia) are used by bulk carriers, and stevedores have to load and unload the cargoes. The possibility of a stevedore being already infected with Ebola before boarding a ship is very real. There is also a problem with stowaways in west African ports. A ship has to be thoroughly searched to ensure that there are no stowaways, possibly infected with Ebola, before a ship can leave port. Ship owners have a duty of care to the crews on their ships, and measures have been put in place to protect them, "to educate the crew about he risks (of Ebola); to check the ISPS regimes on board; to review shore leave and to avoid crew changes in the affected areas". The International Maritime Organisation has an ISPS code; the International Ship and Port Facility Codes that are a set of measures to protect ships and port facilities, put in place after the 9/11 attacks on the US.

Ships owners and captains also have to deal with the refusal of pilots to board a ship that has come from an infected area; or ports being closed to ships from infected areas. There may also be legal claims if a cargo is late arriving after problems in a port affected by Ebola; cargoes may be damaged by late delivery. The cargo ship 'Ocean Discovery' was not allowed to dock at the port of Chaguaramas in Trinidad in October 2014. The ship's last two ports of call were Sekondi Takoradi, Ghana and Pointe Noire, Republic of Congo. A Trinidadian official said, "Ghana is fairly near to Liberia, Ghana and Sierra Leone." Trinidad insisted that the ship had to obey a 45 day incubation period before being allowed to dock. The incubation period for Ebola is 21 days.

The cruise ship 'Magic' was not allowed to dock in Belize or Mexico, because a health worker from the Texas Health Presbyterian Hospital in Dallas was a passenger. The woman had handled samples from Thomas Duncan, who had died of Ebola on 8th October. The health worker was quarantined on the ship, and had not displayed any symptoms of Ebola. Texas Presbyterian announced that it was issuing guidelines limiting travel for staff. Workers would not be allowed to board planes, ships, or trains for 21 days after entering a room which contained a patient suffering from Ebola. The CDC requested that the health worker be sent home from Belize; the Belize government refused as it would not permit the woman to set foot on its country. When the ship arrived back Carnival offered each passenger $200 in ship account credit in compensation and a discount on their next cruise. Most passengers were happy with this: some said they were going to book a new cruise anyway, as they'd had such a good time.

Jamaica, Colombia, Guyana and St Lucia are also refusing entry to travellers who have recently visited Ebola-stricken countries.

Meanwhile, it was announced in London that insurance brokers, in collaboration with Lloyds underwriters, had created quarantine insurance to cover loss of revenue from mandatory quarantine.

The effect of Ebola on animal populations of West Africa

The Ebola outbreak has not only caused devastation to the human population of West Africa, but also to the animals who live there. Since Ebola first appeared in the 1970s, it is estimated that one third of the world's chimpanzees and gorillas have ben killed by the virus. In 1994 alone, the Ebola outbreak in Minkebe, Gabon, "wiped out the entire populations of what used to be the second largest protected population of chimpanzees and gorillas in the world". The protection was against humans - poachers and hunters, nothing could protect the animals from Ebola.

Scientists questioned whether domesticated animals like cats and dogs could be infected with the Ebola virus. During the 2001-2002 outbreak in Gabon, dogs were seen eating the carcasses of dead animals. In African villages dogs are not fed by the villagers, and have to scavenge for their food, eating small dead animals, and entrails of animals being butchered for their meat by hunters. Eric Leroy and is colleagues from the Centre International de Recherches Medicales de Franceville in Gabon, conducted a survey of pet dogs in an area which Ebola had been active. They found that dogs had been exposed to the Ebola virus, but they did not develop symptoms. However, as the dogs lived in close proximity to humans, they can be viewed as a possible risk factor for Ebola infection.

Leroy and his colleagues also observed that Ebola virus is probably secreted in faeces, saliva and urine, when an animal, or human, is infected. Other animals can develop mild symptoms of Ebola, like horses, guinea pigs and goats. The team also suggested that dogs could have been exposed to aerosol exposure, as some were tested in Libreville and Port Gentil, who had never had contact with dead animals, but still demonstrated exposure to Ebola in their blood.

In the UK, the population was assured that, "British dogs will not be put down if their owners get Ebola". The Daily Telegraph stated that there had been widespread outrage over the euthanasing of Teresa Romero's dog in Spain after she was infected with the virus. If Briths people did become infected, their dogs and other pets would be quarantined.

The American Veterinary Medical Association issued guidance that animals had not been responsible for the Ebola outbreak in West Africa, and that if pets came into contact with Ebola, they would be quarantined, and if infected, would be put down.

Ebola is found in other African animals, including duikers and rodents. Pigs, guinea pigs, horses and goats have been infected experimentally and either had no symptoms or mild ones. Ebola has not been found in any African felines, such as lions, so cats may be immune.

Ebola 2014: Forgotten victims

In all the grim figures and statistics about the Ebola outbreak, had anybody counted the number of dead babies? The mortality rate for unborn and newborn babies is close to 100%. Pregnant women are at greatest risk than the general population, and the mortality rate for them is 93%. Pregnant women are much more likely to have a miscarriage or a premature labour if they contract Ebola, and even if they survive, the baby is almost certainly going to die.

The risk of being infected is much higher for nurses and midwives as they come into contact with highly contagious bodily fluids. It is not always possible to have a health worker with experience of delivering babies on hand to help with pregnant women. It is advised that there should be no episiotomies, as the risk of needle-stick is too great. Even delivering a placenta is fraught with danger for health workers, because of the risk of contact with fluids.

Liberia and Guinea were two of the countries with he highest maternal mortality rates in the world before the Ebola outbreak struck. It is estimated that 800,000 womcn in Guinea, Liberia and Sierra Leone will be pregnant by the end of

2015, and the UN Population Fund calculates that around 120,000 of them will have complications. It is not the loss of a woman and her unborn child, dreadful though that is, but the loss of mothers for the unknown number of children in families that are already devastated by the loss of family members to Ebola.

Women are not only mothers, they are also nurses to their family's sick; caregivers to young and old; food providers and cooks. Women make up 51% of the population of Sierra Leone, but they make up 60% of Ebola's fatalities. This figure is the same across the countries affected by Ebola. Women have selflessly cared for the sick; tended the dying; and prepared the dead for burial. It is because of all these practices that women are most affected by Ebola.

One of the few benefits to result from the Ebola outbreak is that the practice of female genital mutilation has gone into decline. In Sierra Leone, about 80% of girls were mutilated, but since Ebola arrived there are very few such procedures. The UN General Assembly passed a resolution in 2012 calling for a worldwide ban, but the practice has continued. There are fears that the mutilator, the sowei, could be infected if the girl already has Ebola, and that the virus could then be passed from girl to girl. Young girls are usually mutilated between the ages of 9 and 12, with the intention of ensuring that they remain virgins until they marry, and are faithful to their husbands. This barbaric procedure, that removes the external genitalia, has been carried out on an estimated 129 million women and girls worldwide, and can cause health complications, even death. Sierra Leone has banned the practice, with a fine of 500,000 leones (about $115) for anybody practicing FGM. The soweis have turned to other occupations to support themselves, and hopefully, when the Ebola outbreak is over, female genital mutilation will become an outdated practice that belongs in the past.

The rising number of children left orphaned by Ebola is horrific. Neighbours do not want to help children for fear of catching the virus, they are often left, alone and helpless, to fend for themselves. At least 20,000 children have ben orphaned in Liberia, and while some charities like Street Child are reaching out to them, many have just been abandoned. There is no government help, and where previously relatives would have taken parentless children in, the fear of catching the virus prevents families doing this.

The cost of food has rocketed in west Africa. Rice is the staple food in Sierra Leone, and most of it is grown on small family farms. If people cannot get to their fields because of quarantine, or family members have died, the rice cannot be planted, or harvested. Even if farmers have been able to grow and harvest rice or other crops, the markets have been closed to prevent infection, so they have been unable to sell their produce. It is a similar situation in Guinea and Liberia, and the Food and Agriculture Organisation of the UN, FAO, is predicting that more than one million people will be facing food insecurity by March 2015. Measures have to be put in place to feed all the people left destitute by Ebola.

Other countries in the region are also being negatively affected by the Ebola outbreak. The Gambia is not affected by Ebola, but tourists are not visiting as they used to. The Gambia is heavily reliant on tourists, and visitors and holidaymakers are staying away - the numbers are 60% down from what they used to be before Ebola. It is not only local people who are affected, but also charities like the Gambia Horse and Donkey Trust. Families rely on horses and donkeys for farming and transportation.

If the animals become ill, the people often cannot afford to pay a vet: a healthy working horse or donkey can increase a farmer's income by up to 500%, according to the Trust. The rainfall was less in 2014, which has had a knock-on effect as grass has not been growing as lushly as in previous years, so fodder for animals is also in short supply.

The Economist stated that the Ebola virus has claimed more victims: African tourism and football. Morocco refused to host the African Cup of Nations in January 2015, citing fears of Ebola in large crowds. In retaliation, the Confederation of African Football banned Morocco, and the event was held in Equatorial Guinea instead: Ivory Coast won. Tourism in Africa has suffered from the Ebola outbreak as outlined by the Economist, "Directly and indirectly, tourism accounts for almost 10% of sub-Saharan Africa's GDP and pays the salaries of millions of people. The industry is worth about $170 billion a year. In 2013 more than 36 million people visited Africa, a figure that has been growing by 6% per year. Now many safari lodges are closer to extinction than the animals around them. Redundant workers might eventually turn to poaching". Most of the countries affected are not anywhere near the Ebola outbreak:

it is 'The ignorance epidemic'.

Incidents of violence

For the first time, Ebola has appeared in heavily populated areas in Guinea, Sierra Leone and Liberia. Fear of the virus has caused violence to flare as rumours and misinformation have spread throughout the crowded cities.

In Freetown, the capital of Sierra Leone, families forcibly removed Ebola patients from hospitals, fearing that the treatment in hospital would kill their loved ones. The fear of hospitals used to be prevalent in richer countries in the past, when medical treatment had to be paid for, and relatives were only admitted for hospital care when an illness was in an advanced stage, and the person concerned was more likely to die. To poorer people, a hospital was where you went to die. A similar fear pervades West Africa: some people would prefer to be treated by a traditional healer; somebody known and trusted, using local herbs and local knowledge, not an apparition in white protective gear.

Violence has also erupted when medical teams have attempted to remove infected people from their homes to take them to treatment centres. In Sierra Leone, people believed that Ebola was a myth invented by the government to allow the authorities to steal their organs or their money.

These reactions were not new. In 2002 in Mbomo, Republic of Congo, a patient was taken from a clinic by a family who preferred treatment by a traditional healer. The patient died, but the healer said that the man had been killed by witchcraft, and 'pointed the finger' at the man's brother who lived in another village. The brother was a schoolteacher, and educated man, so it may have been a simple case of jealousy. A second brother died of Ebola, and a nephew. After these deaths, the teacher's house was burned down, but he managed to escape.

In another case, there was a secret society called the Rose Croix. Four schoolteachers were members. They taught their pupils about Ebola, and how to avoid catching it, but then the virus began killing people. Traditional healers decided that the teachers were sorcerers, and that they had used witchcraft to afflict people. As a result, the local people became enraged, and hacked the teachers to death. When so many people caught the virus, it was decided by the traditional healers that it was not witchcraft after all, but an epidemic.

On 4 April 2014, an angry mob attacked

the Medicins sans Frontieres treatment facility
in Macenta as they thought that the MSF team were introducing Ebola into the community.

In September 2014, a Guinea health team was killed by villagers scared by their appearance. Eight bodies were found, hacked to death by machetes, and dumped in a septic tank near the village of Nzerekore.

Also in September, Red Cross volunteers had buried a body in the mining town Forecariah, in western Guinea, when they were attacked by an armed mob. Two volunteers were seriously injured, and the mob then dug up the body and hid it in the town. At that time, Forecrariah had a case fatality rate of 80% among the townspeople. The mob grew to more than 3000 heavily armed youths, who then attacked a vehicle carrying epidemiologists from the WHO, who fled for safety. All the equipment and vehicles were stolen or vandalised.

In another incident in Sierra Leone, a health team disappeared after being stoned by villagers. When their bodies were found, they had been attacked with clubs and machetes. The villagers thought that health workers were spreading the disease.

A team disinfecting a market at Nzerekore in Guinea, were suspected of deliberately spreading Ebola, and a riot broke out.

In September 2014, it was reported that a delegation of health workers and journalists had arrived at the village of Womey, in south east Guinea, to educate the people about Ebola. At first the delegation was welcomed, but some angry youths arrived and started to stone them. The members of the delegation were later found with their throats cut.

The situation in Sierra Leone was very unsettled even before Ebola arrived. After the decades of civil war, the World Bank estimated that around 60% of the country's young people were unemployed.There were fears not only about Ebola, but that international aid intended to support poorer communities being mismanaged by those in positions of power. The local markets have been affected by Ebola as most people stay in their homes. In October 2014, in Koidu, a political leader refused to allow his sick mother to be taken to an Ebola treatment centre. Youths began to protest, and violence escalated with the police, military, and some of the youths exchanging gunfire.

Some people were killed in the crossfire.

Armed bandits stopped a taxi and stole blood samples believed to be infected with Ebola. The samples were in transit between central Koukan and a testing centre in Gueckedou. The authorities appealed on the radio for the samples to be returned. The bandits obviously did not realise what the flasks contained.

There have been numerous reports of looting from buildings and medical treatment centres. A riot in the West Point area of Monrovia, Liberia, was followed by looting of the clinic. Surprisingly, bedding and blankets covered in blood were stolen. Even a blanket or mattress covered in blood is better than having nothing for people who have no possessions. Can Ebola be caught from an infected blanket? It is not thought to be very likely, but the risk is there, especially if the items are not thoroughly cleaned.

In December, the Assessment Capabilities Project published a report into why there was such resistance to outsiders, and such violence. ACAPS found that some young men were former child soldiers, who had received little or no education during the years of the civil war and unrest in Guinea,

Liberia and Sierra Leone. The lack of response from the authorities had also angered the population: in some places, bodies were left for weeks in the streets because people were too afraid to touch them.

Contact tracing lists were seen by some people as 'death lists', and people refused to cooperate with the authorities. The combination of fear, ignorance, and mistrust led to a constant concern for the health workers trying to help stop the Ebola outbreak.

Health workers on the front line

The Doctor

Dr Javid Abdelmoneim left Kings College Hospital in London to join the fight against Ebola. He agreed to be filmed by the BBC Panorama team, while he was working in Keilehan, Sierra Leone , treating victims of Ebola. He found it very hard, and very tiring. Having to wear hot protective clothing while working inside the treatment tents. It was also emotionally draining, knowing that whatever he did, that people would die. In the programme, it was obvious that Dr Abdelmoneim was growing attached to two one-year-olds, Alpha and Warrah, who had lost their parents to Ebola, and were being cared for by their six-year-old brother, Abdul. Sadly, Abdul died, and a carer had to be found to look after the children, and a lovely lady named Kadia agreed to take on the task. So many people have lost their entire families, wives, husbands, children, to Ebola. Then little Warrah became ill, and died. Dr Javid went with his body to the burial site where so many people were interred. He was visibly upset - the little boy had touched his heart. On his return to the UK, Dr Javid had to monitor his health and temperature for 21 days, but he was fine, he didn't develop

Ebola.

The nurse

Australian Red Cross nurse Sue Ellen Kovack, described her work in the treatment centre in Kenema, Sierra Leone, for the Guardian newspaper. This is a precis of what she said:

At the start of every day, she checks her hands for injuries that would prevent her entering the high-risk zone. Entering the centre, the hands are washed in 0.5% chlorine solution, and somebody else sprays the bottoms of the shoes with the solution. In the low-risk area, cold, wet boots are waiting that have been soaking in chlorine overnight. Then she dresses in scrubs, and enters the low-risk area. A whiteboard gives information about who has died during the night: one of the nurses has died. A dresser makes sure that everyone is covered with PPE (personal protective equipment), and they check each other to make sure that every part of the body is completely covered.

First on is a jumpsuit and gloves. Then a second pair of gloves, a duckbill mask, a hood, and an apron. The apron is specially tied so that it can be undone with one pull.

Then the mask, and a final check to make sure that not one inch of skin is exposed.

Then into the high-risk zone. Time inside is only about an hour, due to the necessity of not over-heating inside the PPEs: if a nurse or other healthcare worker faints, it is very risky for their colleagues to try and get them outside.

Some patients are too ill to move away from their own vomit, faeces and urine, and the healthcare workers clean them up, and ensure that they are as comfortable as possible. Patients who are getting better also have to be looked after with food, fluids, cleaning and pain relief.

At the end of an hour, the team go into the undressing area. They have to be sprayed all over with chlorine, especially the hands. Then, feet apart and hands in the air they are sprayed all over again, front and back.The hands are washed in chlorine, then the first pair of gloves is removed. Then the apron comes off, and goes straight into a bath of chlorine. The hands are washed again. Then the goggles are very carefully removed, "We bend over, close our eyes and gently remove them, dunk them three times in thc strong chlorine-filled bucket, and then place them in water." The hood is removed in a similar way, then the heavy PPE is removed, very slowly to avoid accidentally touching any of the outside of it. Then the jumpsuit is kicked off, and placed in chlorine. Finally, the heavy-duty filtration mask is removed. The hands are washed again, then the last pair of gloves is removed. The boots are sprayed with chlorine from all angles, and the job is done.

The move from high-risk to the low-risk area comes last. As Sue Ellen Kovack says, "I need rehydration with water. No food is allowed in the low-risk area. It is too risky to put anything near your mouth from your hands. But I still see people biting their nails, touching their face, rubbing their eyes - risky but automatic responses. Your hands have been washed a trillion times in chlorine, but still, ...Before I left Australia, I took to wearing a rubber band and each time I caught myself touching my face, I snapped it painfully so I would remember not to do it".

The WASH (water, sanitation and hygiene) staff

For every one doctor and three nurses working at an Ebola treatment centre, there are approximately twenty-six WASH staff.

They are recruited locally, and some are teachers who lost their jobs when the schools closed. The WASH staff are responsible for cleaning and disinfecting treatment wards,;cleaning spills of blood, faeces, vomit and urine; emptying buckets of waste fluids; handling dead bodies; and operating the incinerators. For this they are paid very little. They wear full personal protective equipment, and wear large green gloves: the nursing staff wear small white ones.

The people on the WASH team often do not tell their families what they are doing. It is dirty, and very dangerous work. There is stigma involved as the local community shun people who work with Ebola. "It takes bravery to leave one's community to travel overseas to work in an Ebola treatment unit. But it takes extraordinary bravery to work in an Ebola treatment unit without leaving one's community".

Body management team

Melvin Gibson works for the Liberian Red Cross Body Management Team. He said, "It was so difficult. I felt so discouraged. My very first day, we picked up 52 bodies. I was picking up my friends - Liberians. I did it because at the time nobody was willing to do this work. And this is my country".

The men describe how difficult it is to persuade the relatives to allow a body to be taken for burial. People do not like to admit that their relative died of Ebola. It is so very difficult.

Mark Korvayan, who works on a burial team in Liberia, has been attacked many times. Local people often do not want a 'strange' body buried in their local cemetery, especially a body of someone who has died of Ebola, which has such stigma attached to it. The police, and often the army, escort the burial teams in case of trouble.

Handling Ebola bodies means wearing personal protective equipment, as often the bodies are leaking, and they are full of Ebola virus. The teams have to take their protection off every 45 minutes as it gets too hot to work in it.

Sometimes, the collection of a body is touching: when the body of Shettema Bole, aged 45, the wife of Tamba Bole, is removed, the watching neighbours sing, "Where are you going dear mother, why are you leaving us so young?" The removal was dignified and solemn.

Time magazine voted The Ebola Fighters, Person of the Year 2014, for the harrowing, and very brave work that the nurses, doctors, WASH teams, body management teams, and all those involved in caring for those sick, or dead, of Ebola, in all the affected countries.

The Ebola outbreak in 2015: what does the future hold for West Africa?

2015 began with good news; Ebola infections were slowing, and it seemed that the epidemic might be under control. In Liberia, the government lifted the curfew for one night so that people could go to church to celebrate the dawn of a new year.

West Point is a shanty town in Monrovia, the capital of Liberia. The government had feared that Ebola would be a threat to the entire community, but although there had been many cases, the massive infections that were feared have not happened. At the Bethel World Outreach church, an usher armed with a radar gun thermometer, checked everybody entering; anyone with a raised temperature could be infected with Ebola. The Liberians say, "Happy New Year, Me No Die Oh", to each other on New Year's eve, and the celebrants at the church were upbeat about what the future holds for them.

In Liberia, schools are still closed, and many health workers heave died, but new treatment centres have opened with newly trained health workers, and nine laboratories have been deployed across the country for the early detection of Ebola.

People are beginning to feel more positive about the future.

The situation in Guinea is less positive as there have been an increasing number of cases around Conakry.:74 new cases during the first week of January. Sierra Leone's new cases are also giving cause for concern: 250 new cases in the same week. It is thought that the problem lies in the rural areas, where people are still hiding infected people, and still persisting with practices like washing corpses before burial. Dr Jeremy Ferrar, director of the Wellcome Trust said that, "All the time the epidemic smoulders, it runs the risk of reigniting."

One of the main reasons that diseases like SARS and smallpox were successfully contained or eradicated was in contact tracing. Anybody who had been in contact with an infected person had to be traced and quarantined. This system just has not worked in some parts of West Africa with the Ebola outbreak. People in Africa often do not have an address, because people move around a lot and often take the front door with them. Doors cost money and poorer people cannot afford to buy a new front door every time they move.

People are often very suspicious of health workers, especially if they come from outside the area.

Health support workers are often working with outdated methods, such as using Excel spreadsheets, when the software designed by the CDC can be carried on a thumb drive and used wherever it is needed. Once workers were trained in the new methods, organising thousands of names in a fraction of the time, and supplied with new computers, tracing contacts began to improve.

There have been problems tracing contacts across borders, especially in rural areas where villages are scattered with poor road access. In Guinea, village 'watch committees' were proposed in an effort to engage community leaders, who would get the local population involved. By December 2014, only half of the proposed committees had been set up and were functioning.

During the final few months of 2014, there was a surge in infected healthcare workers, especially in Guinea. An investigation discovered that there was an insufficient supply of personal protective clothing. With the amount of aid from countries across the world, both financial and in personnel, a lack of protective clothing should not have been an issue.

Guinea is divided into prefectures, and in October 2014 there were only seven with reported cases; by the middle of December the number had grown to seventeen. The level of community resistance to control measures had also grown in Guinea. There is still a long way to go before preventive measures are accepted by all of the population.

The WHO has been widely criticised for its slow, and ineffective, response to the Ebola outbreak. At a special session of the WHO executive board on 25 January 2015, a resolution was passed that requested a 'transformation in the way the organisation works in health emergencies'. Ebola has cost nearly 10,000 deaths despite $4 billion in emergency aid from seventy countries.

The WHO has six regional offices around the world, and the failure of the Africa office, and the HQ in Geneva, to deal with the Ebola outbreak quickly enough, has been strongly condemned. It is proposed that the Director General of the WHO, who at the moment is Margaret Chan, should reassert her authority over the whole of the organisation, instead of having six autonomous local heads. The Director General was also instructed to commission an interim assessment of what has gone wrong in the WHO, to be ready to report by May 2015.

The UK has sent 600 military personnel to work in Sierra Leone, and one healthcare worker was airlifted back to the Royal Free Hospital in London on 1 February 2015, with a needle-stick injury. More than a dozen doctors and a hundred healthcare workers have died in Sierra Leone so far. The WHO reported on 25 February 2015 that in Guinea, 3155 people have been infected, of whom 2091 have died. In Liberia, 9238 have been infected, of whom 4037 have died, and in Sierra Leone, 11301 have been infected, of whom 3461 have died. The totals are: 23,729 infected, of whom 9604 have died. The CDC has reported that hundreds of thousands more people could become infected with Ebola, with a predicted 500,000 by the end of January 2016. In the past 21 days, there have been 152 new cases in Guinea; 10 new cases in Liberia, and 235 new cases in Sierra Leone.

In Sierra Leone, there have been 31 new cases in one village - Makeni. The village has been put into lockdown by the army to try and prevent further spread of Ebola. The WHO says that the new cases are linked to one man who escaped from quarantine in Freetown and went home to Makeni to get treatment from a traditional healer. After receiving treatment, the man's father and the healer both died of Ebola.

There have been more attacks on health workers who were trying to advise local people on how to bury a body without being contaminated with Ebola. The people believe that it is the health workers who are bringing Ebola, and they just do not trust them. A lot more effort is needed to educate people about Ebola.

What is more worrying in Guinea, is that the proportion of new cases where the source is unknown is only about 15%. That means that 85% of new cases are from an unknown source: this could mean hat Ebola is becoming endemic in areas that are high-risk, and will not be eradicated.

The governments of West Africa are stressing that here are business opportunities for companies brave enough to invest in the area. Guinea, Liberia and Sierra Leone have each lost more that 10% of their GDP as a result of Ebola, and governments are very keen to attract new businesses.

It is estimated that a further $1.5 billion will be needed to fight Ebola in the first half of 2015. There are hot spots of infection deep in the rain forest, and until contact tracers and burial teams can work in these regions, Ebola will not be stopped. The Ebola virus is both terrifying, and fascinating. A vaccine will be developed that will cure the disease, but there are still the unanswered question: where does Ebola come from? Where does it hide? It will conceal itself again in the rain forest, until the next time it emerges to infect, kill and terrify the human population of Africa and the world.

Select bibliography

Books

Close, William T. 1995. *Ebola*. London: Arrow.

Desowitz, Robert. 1997. *Tropical diseases from 50000 BC to 2500 AD*. London: HarperCollins

Gadarowski, Jennifer. 2014. *Ebola: from bioweapon to pandemic*. Brain Feud

Horowitz, Leonard G. 2014. *Emerging viruses: AIDS & Ebola: nature, accident or intentional?*. London: Tetrahedron

McCormick, Joseph B & Fisher-Hoch, Susan. 1996. *Level 4: virus hunters of the CDC.*
New York: Barnes & Noble

Moore, Peter. 2009. *Pandemics: 50 of the world's worst plagues and infectious diseases.*
London: New Holland

Oldstone, Michael. 1998. *Viruses, plagues, and history.* Oxford: Oxford University Press

Peters, CJ. 1998. *Virus hunter: thirty years of battling hot viruses around the world.* New York: Anchor

Preston, Richard. 1995. *The hot zone.* London: Corgi

Quammen, David. 2014. *Ebola: the natural and human history.* London: Bodley Head

Quammen, David. 2013. *Spillover: animal infections and the next human pandemic.* London: Vintage

Regis, Ed. 1996. *Virus Ground Zero: stalking the killer viruses with the Centers for Disease Control* . New York: Pocket

Regush, Nicholas 2001. *The virus within: a coming epidemic.* London: Vision

Revill, Jo. 2005. *Everything you need to know about Bird Flu & what you can do to protect yourself.* London: Rodale

Ryan, Frank. 1998. *Virus X: understanding the real threat of the new pandemic.* London: HarperCollins

Wolfe, Nathan. 2011. *The viral storm: the dawn of a new pandemic.* London: Allen Lane

Journal articles

Adams, Mike. 2014 Why do the US Centers for Disease Control own a patent on Ebola 'invention'? *Global Research* 12 August

Formenty, P ...[et al] 1999. Ebola virus outbreak among wild chimpanzees living in a rain forest of Cote d'Ivoire. *Journal of Infectious Diseases.* February. Supplement 1. p.120-126

Hayman, David TS ...[et al] 2012. Fruit bats are the presumptive reservoir hosts of Ebola Virus. *Emerging Infectious Diseases* 18(7), p.1207-1209

Leroy, GM., Gonzalez, JP., Baize, S. 2011. Ebola and Marburg haemorrhagic fever viruses: major scientific advances, but a relatively minor health threat for Africa. *Clinical Microbiological Infection.* 17(7). p.964-967

MacNeil, Adam ...[et al]. 2010. Proportion of deaths and clinical features in Bundibugyo Ebola virus infection, Uganda. *Emerging Infectious Diseases.* 16(12). P.1969-1972

Papagrigorakis, MJ ...[et al]. 2006. DNA examination of ancient dental pulp incriminates typhoid fever as probable cause of the Plague of Athens. *International Journal of Infectious Diseases.* 10. p.206-214

Report of an International Commission. 1978. Ebola haemorrhagic fever in Zaire 1976. *Bulletin of the World Health Organisation*. 56(2). p.271-293

Taylor, DJ., Leach, RW., Bruenn, J. 2010. Filoviruses are ancient and integrated into mammalian genomes. *BMC Evolutionary Biology*. 22(10), p.193

Walsh, Peter, Real, Leslie, Biek, Roman. 2005. Wave-like spread of Ebola Zaire. *PLoS Biology.* 3(11). p371

Wittman, TJ ...[et al]. 2007. Isolates of Zaire ebolavirus from wild apes reveal genetic lineage and recombinants. *Proceedings National Academy Sciences USA.* 104(43). p. 17123-17127

Media

Abdallah, Dodi. We will run from Ebola patients. *Ghana Web.* 2 August 2014

Allen, Vanessa. Angry nurses pelt Spain PM with rubber gloves. *Daily Mail*. 11 October 2014

Allen, Vanessa. Panic on plane as man shouts: I've got Ebola! *Daily Mail*. 11 October 2014

Associated Press, Conakry. Bandits in Guinea steal blood samples believed to be infected with Ebola. *The Guardian*. 21 November 2014

Baker, Aryn. Racing Ebola: what the world needs to do to stop the deadly virus. *Time,* 13 October 2014

Baker, Aryn. The race to diagnose: better and faster Ebola testing in West Africa would save lives and could help bring an end to the outbreak. *Time*. 3 November 2014

Bassant, Mark. Ship coming from Africa refused entry. *Trinidad Express*. 16 October 2014

Boseley, Sarah. Ebola endemic in West Africa remains a risk, scientists warn. *The Guardian*. 25 February 2015

Boseley, Sarah. One in seven pregnant women could die in Ebola-hit countries, say charities. *The Guardian*. 10 November 2014

Brown, Rob. The virus detective who discovered Ebola in 1976. *BBC News Magazine*. 18 July 2014

CDC Case Counts. 27 January 2015

CDC Ebola guidance for airlines. 15 October 2014

Cleric: use of WMD by terrorists against non-Muslims violates Islamic law. *The Jerusalem Post*. 4 February 2015

Calzac, Natasha. US missionary 'idiotic and narcissistic'. *The Independent.* 8 August 2014

Dart, Tom. Tourists back in Texas from cruise ship carrying woman monitored for Ebola. *The Guardian*. 19 October 2014

Dawber, Alistair. Ebola outbreak: disease ebbs in West Africa as aid agencies warn against complacency. *The Independent*. 28 February 2015

Devries, Nina. Ebola fears bring female genital mutilation to near halt in Sierra Leone. *Al Jazeera*. 4 December 2014

Doornbos, Harold. Found: the Islamic State's terror laptop of doom. *Foreign Policy*. 28 August 2014

Ebola can be turned into a bioweapon, Russian & UK experts warn. *Russia Today*. 8 August 2014

Freman, Colin. Doctor sues Spanish nurse over claim that she prescribed her paracetamol. *Daily Telegraph*. 28 November 2014

Gallagher, James. 5 ways to avoid catching Ebola. *The Observer*. 19 October 2014

Hamilton, Lisa M. Ebola's hidden costs. *The Atlantic*. 14 January 2015

Harding, Andrew. Ebola crisis: a village fights back in Sierra Leone. *BBC News*. 4 November 2014

Hicks, Edward, and Sandhu, Jasmine. Ebola virus: the legal challenges. *Law Gazette*. 29 September 2014

Horizon. Ebola: the search for a cure. *BBC 2*. 9 October 2014

Kirby, Alex. Animals may give Ebola warning. *BBC News*. 15 January 2015

Leaung, Marlene. Worst kind of flu you can imagine. *CTV News, Canada*. 5 August 2014

McVeigh, Tracy. NHS staff 'insulted' by UK travel ban. *The Observer*. 21 December 2014

McVeigh, Tracy. Oxfam in call for troops as criticism of 'inadequate' Ebola response mounts. *The Observer*. 19 October 2014

Medicins sans Frontieres. Ebola crisis update. 21 November 2014

Nebehay, Stephanie. Ebola likely to persist in 2015 as communities resist aid. *Reuters*. 30 January 2015

O'Carroll, Lisa. Ebola: Sierra Leone village in lockdown after 31 new cases recorded. *The Guardian*. 27 February 2015

Park, Alice. What's the status of all the experimental drugs and vaccines? *Time*. 3 November 2014

Piot, Peter. In 1976 I discovered Ebola. Now I fear an unimaginable tragedy. *The Observer*. 5 October 2014

Ramirez, Anthony. Was the Plague of Athens really Ebola? *The New York Times*. 18 August 2014

Rush, James. Summit in London to discuss global response. *The Independent*. 2 October 2014

Schoepp, Randal J. Sierra Leone samples: Ebola evidence in west Africa in 2006: USAMRIID providing on-site laboratory support to current outbreak. *USAMRIID, Fort Detrick*. 14 July 2014

Selby, Jenn. Ebola doctors must 'suffer the consequences'. *The Independent*. 4 August 2014

Shipping issues arising out of the Ebola outbreak. *Steamship Manual*. 9 January 2015

Stylianou, Nassos. How world's worst Ebola outbreak began with one boy's death. *BBC News*. 27 November 2014

Tekmira establishes manufacturing and clinical trial agreement to provide TKM_Ebola-Guinea for clinical studies in west Africa. *Tekmira Global News* 22 December 2014

Thompson, Ken. The toxic charms of nicotiana. *The Daily Telegraph*. 12 August 2014

Vlahos, Kelley Beaucar. How a Virginia suburb became an Ebola epicenter. *Fox News*. 19 September 2014

WHO Fact sheet no. 103. Updated April 2014. Ebola virus disease

WHO. Known cases and outbreaks of Ebola virus disease in reverse chronological order

WHO Situation report. 7 January 2015

WHO Situation report. 25 February 2015

18470117R00080

Printed in Great Britain
by Amazon